Fabian Grein

The DsrMKJOP transmembrane complex from Allochromatium vinosum

Fabian Grein

The DsrMKJOP transmembrane complex from Allochromatium vinosum

- a biochemical, biophysical and functional analysis

Südwestdeutscher Verlag für Hochschulschriften

Imprint
Any brand names and product names mentioned in this book are subject to trademark, brand or patent protection and are trademarks or registered trademarks of their respective holders. The use of brand names, product names, common names, trade names, product descriptions etc. even without a particular marking in this work is in no way to be construed to mean that such names may be regarded as unrestricted in respect of trademark and brand protection legislation and could thus be used by anyone.

Publisher:
Südwestdeutscher Verlag für Hochschulschriften
is a trademark of
Dodo Books Indian Ocean Ltd., member of the OmniScriptum S.R.L Publishing group
str. A.Russo 15, of. 61, Chisinau-2068, Republic of Moldova Europe
Printed at: see last page
ISBN: 978-3-8381-2669-2

Zugl. / Approved by: Bonn, Rheinische Friedrich-Wilhelms-Universität, Diss., 2010

Copyright © Fabian Grein
Copyright © 2011 Dodo Books Indian Ocean Ltd., member of the OmniScriptum S.R.L Publishing group

CONTENTS

1	INTRODUCTION	1
1.1	Sulfur Metabolism	1
1.2	*Allochromatium vinosum*	2
1.3	The *dsr* operon and cytoplasmic Dsr proteins	2
1.4	The DsrMKJOP transmembrane complex	5
1.5	Hemes and iron-sulfur clusters	8
1.6	Aim of this work	12
2	**METHODS**	**13**
2.1	Bacterial strains, antibodies, plasmids and oligonucleotides	13
2.2	Chemicals, enzymes and kits	16
2.2.1	Chemicals	16
2.2.2	Enzymes	16
2.2.3	Kits	16
2.3	Software and online tools	17
2.4	Other materials	17
2.5	Microbiological methods	18
2.5.1	Media	18
2.5.2	Antibiotics	21
2.5.3	Cultivation of *A. vinosum* and *E. coli*	22
2.5.4	Conservation	23
2.5.5	Preparation of competent *E. coli* cells	23
2.6	Analytical methods	23
2.6.1	Determination of protein concentration	23
2.6.2	Heme quantification	24
2.6.3	Quantification of iron sulfur clusters	24
2.6.4	Determination of elemental sulfur	25

	2.6.5	Determination of thiols by HPLC	25
	2.6.6	Determination of sulfate by HPLC	26
2.7	**Molecular biological methods**		**27**
	2.7.1	Polymerase chain reaction (PCR)	27
	2.7.2	GeneSOEING	27
	2.7.3	Enzymatic DNA modification	27
	2.7.4	DNA visualization	28
	2.7.5	DNA preparation and purification	28
	2.7.6	DNA sequencing	29
	2.7.7	DNA transfer	30
	2.7.8	Southern blotting	31
2.8	**Working under anoxic conditions**		**32**
2.9	**Proteinbiochemical methods**		**32**
	2.9.1	Cell lysis	32
	2.9.2	Ultracentrifugation	33
	2.9.3	Solubilization of membrane proteins	33
	2.9.4	Chromatography methods	33
	2.9.5	Electrophoretic protein separation	34
	2.9.6	Protein visualization and identification	36
	2.9.7	Western blotting	36
	2.9.8	Immunological protein detection	37
	2.9.9	Ponceau staining	38
	2.9.10	Buffer exchange and protein concentration	38
	2.9.11	Analysis of protein interactions by coelution assays	38
	2.9.12	Quinone interaction	39
	2.9.13	*In vitro* reconstitution of FeS clusters	39
	2.9.14	Mass spectrometry	39
2.10	**Protein biophysical methods**		**39**
	2.10.1	UV/Vis absorption spectroscopy	40
	2.10.2	Redox titrations	41
	2.10.3	EPR spectroscopy	43
	2.10.4	Resonance Raman spectroscopy	45
2.11	**Bioinformatic methods**		**46**
3	**RESULTS**		**47**

3.1	Enrichment and initial characterization of DsrMKJOP from *A. vinosum*	47
3.1.1	Construction of mutants for the enrichment of DsrMKJOP	47
3.1.2	Purification of DsrMKJOP	48
3.1.3	Characterization of DsrMKJOP	49
3.2	**Individual production and characterization of DsrJ**	**56**
3.2.1	Bioinformatic analysis of DsrJ	56
3.2.2	Heterologous production of recombinant DsrJ	56
3.2.3	Purification of DsrJ and DsrJC46S	58
3.2.4	Characterization of recombinant DsrJ and DsrJC46S	60
3.3	**Investigating the role of cysteine 46 *in vivo***	**71**
3.4	**Individual production and characterization of DsrM**	**73**
3.4.1	Heterologous production of recombinant DsrM	73
3.4.2	Purification of DsrM	74
3.4.3	Characterization of DsrM	75
3.5	**Individual production and characterization of DsrP**	**80**
3.5.1	Heterologous production of recombinant DsrP	80
3.5.2	Purification of DsrP	80
3.5.3	Characterization of DsrP	81
3.6	**Individual production and characterization of DsrK**	**85**
3.6.1	Heterologous production of recombinant DsrK and cellular localization	85
3.6.2	Purification of DsrK	86
3.6.3	Characterization of DsrK	86
3.7	**Individual production and initial characterization of DsrO**	**90**
3.7.1	Heterologous production of recombinant DsrO	90
3.7.2	Purification of DsrO	91
3.7.3	Initial characterization of DsrO	91
4	**DISCUSSION**	**93**
5	**OUTLOOK**	**106**
6	**SUMMARY**	**107**
7	**REFERENCE LIST**	**110**

Abbreviations

Amp	ampicillin
APS	ammonium persulfate
BSA	bovine serum albumin
CIAP	calf intestine alkaline phosphatase
CISM	complex iron–sulfur molybdoenzyme
Cm	chloramphenicol
ddH$_2$O	demineralized deionized water
DDM	n-dodecyl-β-D-maltoside
dH$_2$O	demineralized water
dig	digoxigenin
DMSO	dimethylsulfoxide
DNA	deoxyribonucleic acid
E$_0$	midpoint redox potential
EDTA	ethylenediaminetetraacetic acid
EPR	electron paramagnetic resonance
HEPES	4-(2-hydroxyethyl)-1-piperazineethanesulfonic acid
HPLC	high performance liquid chromatography
IPTG	isopropyl-β-D-thiogalactoside
Kan	kanamycin
LC-MS	liquid chromatography-mass spectrometry
LDS	lithium dodecyl sulfate
MWCO	molecular weight cut off
NTA	nitrilotriacetic acid
OD	optical density
PAGE	polyacrylamide gel electrophoresis
PCR	polymerase chain reaction
Rif	rifampicin
RR	resonance Raman
SDS	dodium dodecyl sulfate
TCA	trichloroacetic acid
TEMED	N,N,N',N'-tetramethylendiamine
TMBZ	3,3',5,5'-Tetramethylbenzidine
Tris	trishydroxymethylaminomethane
HDR	heterodisulfide reductase
rt	room temperature
p.a.	per analysis

1 INTRODUCTION

1.1 Sulfur Metabolism

Sulfur is essential for all living organisms. The incorporation of sulfur into organic molecules like the sulfur containing amino acids methionine or cysteine, or into other organic molecules is referred to as assimilatory sulfur metabolism. It is accomplished by several bacteria, archaea, funghi and plants.

In dissimilatory sulfur metabolism, sulfur compounds serve either as electron acceptors or donors to gain energy. This metabolism is restricted to prokaryotes and can be further divided into the reductive and the oxidative dissimilatory sulfur metabolism. In the reductive dissimilatory sulfur metabolism, sulfur compounds such as sulfate, sulfite, thiosulfate or elemental sulfur serve as electron acceptors. The electrons derive from molecular hydrogen or organic compounds. This metabolism is accomplished by a number of δ proteobacteria (with *Desulfovibrio* species being the most prominent members) but also by Gram-positive bacteria (e.g. *Morella thermoacetica*) and archaea (e.g. *Archaeoglobus fulgidus*). In the oxidative dissimilatory sulfur metabolism reduced sulfur compounds or elemental sulfur are oxidized by a number of prokaryotes. Furthermore there are some organisms that can accomplish either the reductive or the oxidative metabolism, depending on the environmental conditions. An example is the genus *Acidianus*. Organisms gaining energy from oxidative sulfur metabolism are either bacteria (e.g. *Paracoccus* sp.) or archaea (e.g. *Acidianus* sp.) growing chemolithoautotrophically or bacteria gaining energy by anoxygenic photosynthesis. Two well known examples of the latter are the green sulfur bacteria (*Chlorobiaceae*) and the purple sulfur bacteria. The purple sulfur bacteria are divided into two families that can be distinguished by the localization of the sulfur globules that are formed as obligate intermediates during sulfur oxidation. While the sulfur globules are localized extracellularly in the *Ectothiorhodospiraceae* the globules are accumulated inside of the cell in the *Chromatiaceae*. *Thiorhodospira sibirica* belongs to the *Ectothiorhodospiraceae* and represents an exception as it can accumulate sulfur globules in or outside of the cell (Bryantseva et al., 1999). A prominent member of the *Chromatiaceae* is *Allochromatium vinosum* that has been studied in this work and that is described in the next section.

1.2 *Allochromatium vinosum*

Allochromatium vinosum is a Gram-negative rod belonging to the family *Chromatiaceae* within the γ proteobacteria. It's typical habitats are stagnant fresh and salt water and sediments containing hydrogen sulfide (Pfennig & Trüper, 1989). The purple sulfur bacterium grows photolithoautotrophically with reduced sulfur compounds (sulfide, polysulfide, thiosulfate, sulfite or elemental sulfur) or with molecular hydrogen as electron donors (Steudel et al., 1990; Imhoff, 2005). As sulfur compounds are oxidized, sulfur globules are formed as obligate intermediates that can be further reduced to sulfate. Energy is obtained via anoxygenic photosynthesis and carbon dioxide is fixed via the reductive pentose phosphate pathway (Brune, 1989). *A. vinosum* is also able to grow photoorganoheterotrophically where organic compounds (e.g. acetate, malate, fumarate, succinate) serve as electron donors (Imhoff, 2005). *A. vinosum* is an ideal model organism for studying the oxidative sulfur metabolism in phototropic purple sulfur bacteria, because methods for genetic manipulation have been established (Pattaragulwanit & Dahl, 1995) and the complete genome sequence is available (NC_013851).

1.3 The *dsr* operon and cytoplasmic Dsr proteins

In the purple sulfur bacterium *Allochromatium vinosum* the degradation of sulfur globules, is strictly dependent on the proteins encoded in the *dsr* operon (Dahl et al., 2005). The *dsr* genes are not only found in sulfur-oxidizing bacteria but also in sulfate- and sulfite-reducing bacteria and archaea although the composition of some of the Dsr proteins differs between sulfate-reducing organisms and sulfur oxidizers. The *dsr* operon of *A. vinosum* consists of 15 genes (*dsrABEFHCMKLJOPNRS*) and their expression is regulated by a sulfide inducible promoter located upstream of *dsrA* (Dahl et al., 2005; Pott & Dahl, 1998; Grimm et al., 2010b). Besides this, additional secondary promoters are proposed for *dsrS* and *dsrC* (Pott & Dahl, 1998; Grimm et al., 2010b).

The first two genes of the *dsr* operon (*dsrAB*) encode the dissimilatory sulfite reductase (dSiR) from which the operon derived its name. The protein from *A. vinosum* is closely related to the dissimilatory sulfite reducases from sulfate-reducing prokaryotes (Hipp et al., 1997). It is thought to act in the reverse direction in sulfur-oxidizing bacteria (Schedel et al., 1979). dSiR from *A. vinosum* is a cytoplasmic heterotetrameric protein containing siro(heme)amide-[4Fe-4S] as prosthetic group (Lübbe et al., 2006). This prosthetic group is the amidated form of siroheme coupled to a [4Fe-4S] cluster via a cysteine axial ligand. DsrN is a homologue of cobyrinic acid *a,c*-diamide synthase and is thought to catalyze the glutamine-dependent amidation of siroheme (Lübbe et al., 2006). The

1 Introduction

structures of dissimilatory sulfite reductases from the two sulfate-reducing organisms *Archaeoglobus fulgidus* and *Desulfovibrio desulfuricans* have been solved (Schiffer et al., 2008; Oliveira et al., 2008). dSiR from *Archaeoglobus fulgidus* has been reported to contain four siroheme-[4Fe-4S] clusters, only two of which are catalytically active (Schiffer et al., 2008). The structure from *D. desulfuricans* revealed that the enzyme contains two sirohydrochlorins (i.e. the demetalled form of siroheme) and two catalytically active siroheme-[4Fe-4S] clusters (Oliveira et al., 2008). The latter structure moreover provides evidence for the direct involvement of DsrC in sulfite reduction (Oliveira et al., 2008). In sulfate-reducing organisms this protein has initially been reported as γ subunit of dSiRs as preparations of DsrAB were found to contain also DsrC (Steuber et al., 1995; Pierik et al., 1992). However, *dsrC* is not located within the same operon as *dsrAB* in several organisms and the expression of *dsrAB* and *dsrC* genes is not coordinately regulated. Furthermore, there are preparations of DsrAB that do not contain DsrC. These facts led to the conclusion that DsrC rather interacts with dSiR instead of being a subunit of this enzyme (Oliveira et al., 2008).

The solution structure of recombinant DsrC from *A. vinosum* has been solved and it revealed a highly conserved C-terminus that forms a flexible arm (Cort et al., 2008). Within this arm, two strictly conserved cysteines are located. These cysteines were found to undergo thiol–disulfide interconversions. DsrC has also been shown to interact with DsrEFH – a protein that is restricted to sulfur-oxidizing bacteria - and the conserved cysteine at the penultimate position of DsrC has been shown to be essential for this interaction (Cort et al., 2008). The structure of DsrEFH from *A. vinosum* has been solved as well and the protein was shown to form an $\alpha_2\beta_2\gamma_2$ heterotrimer. It has furthermore been found that a conserved active site cysteine in DsrE is essential for the interaction with DsrC (Dahl et al., 2008). These results together led to a proposed function in sulfur binding/transferring chemistry from DsrEFH to DsrC. The latter could finally act as a direct or indirect substrate donating protein to dSiR (Cort et al., 2008). An additional function of DsrC as a regulatory protein has been proposed (Grimm et al., 2010b). A putative interaction of DsrEFH, DsrC and DsrAB is corroborated by the fact that they are copurified when DsrAB is purified from the membrane fraction of *A. vinosum*. Furthermore the proteins DsrK, DsrO and DsrJ are copurified, indicating an interaction of the DsrMKJOP transmembrane complex as well (Dahl et al., 2005). This has been taken into account in the model of sulfate-reduction proposed by Oliveira et al. (2008). The DsrMKJOP transmembrane complex is the central topic of this work and will therefore be described in the next section.

1 Introduction

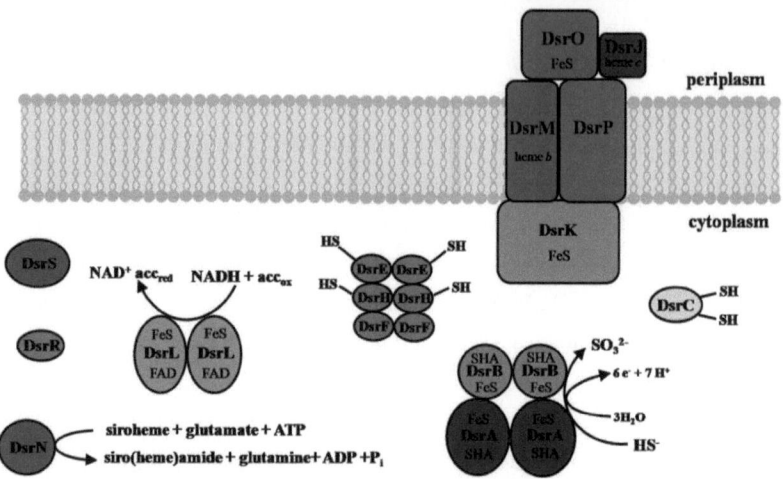

Figure 1.1: Schematic localization of the Dsr proteins; cofactor binding is indicated; SHA, sirohemeamide; Illustration of the DsrMKJOP complex is based on the information available prior to this study.

A. vinosum DsrL is a homodimeric iron-sulfur flavoprotein and is copurified together with DsrAB when the latter is purified from the soluble fraction. The protein exhibits an NADH:acceptor oxidoreductase activity and carries a thioredoxin-like CXXC motif (Lübbe et al., 2006; Lübbe, 2005). Its function is not known yet but it has been proposed that it could be involved in the reductive release of sulfide from a perthiolic organic carrier molecule. This molecule may be the link between the periplasmic sulfur globules and the cytoplasmic dSiR (Dahl et al., 2005). Accordingly DsrL is only found in sulfur-oxidizing bacteria.

The proteins DsrR and DsrS resemble small proteins that reside in the cytoplasm. It has been proposed that DsrR is involved in the posttranscriptional control of the *dsr* operon (Grimm et al., 2010a). The function of DsrS is currently not known.

1.4 The DsrMKJOP transmembrane complex

The importance of the DsrMKJOP transmembrane complex for the sulfur oxidation in *A. vinosum* has been studied in mutants carrying *in frame* deletions of each individual gene. None of the mutants was able to oxidize stored sulfur revealing that the complex has an essential function in sulfur oxidation (Sander et al., 2006). As mentioned above, the complex is also present in sulfate-reducing prokaryotes. Although genes coding for several different transmembrane protein complexes are present in genomes of sulfate-reducing organisms, the genes coding for the DsrMKJOP complex are apparently strictly conserved. This indicates a vital role in sulfate reduction as well (Pereira, 2007).

The DsrMKJOP transmembrane complex was first isolated from the sulfate reducing archaeon *Archaeoglobus fulgidus* as Hme complex (Mander et al., 2002), and afterwards from *D. desulfuricans* (Pires et al., 2006). The DsrMKJOP complex consists of cytoplasmic, membrane integral and periplasmic components, and is predicted to be involved in electron transfer across the membrane (Dahl et al., 2005; Pott & Dahl, 1998; Pires et al., 2006). The individual components of the complex will be described below.

DsrJ

The sequence of DsrJ from *A. vinosum* has been entered in the NCBI database and can be found there under the accession number YP_003443231. The protein shows no sequence similarity to any other protein in the database with the exception of DsrJ proteins from other organisms. The amino acid sequence deduced from the *dsrJ* gene contains three possible CXXCH heme binding motifs (Figure 1.2). In addition, as already pointed out by Pires et al. (2006) for DsrJ from a dissimilatory sulfate reducer, sequence analysis suggests that the distal residues responsible for the axial coordination of the hemes may be histidine, methionine and cysteine, respectively. In total, four conserved residues can be found in the alignment that may act as distal ligands to the three hemes (Figure 1.2).

1 Introduction

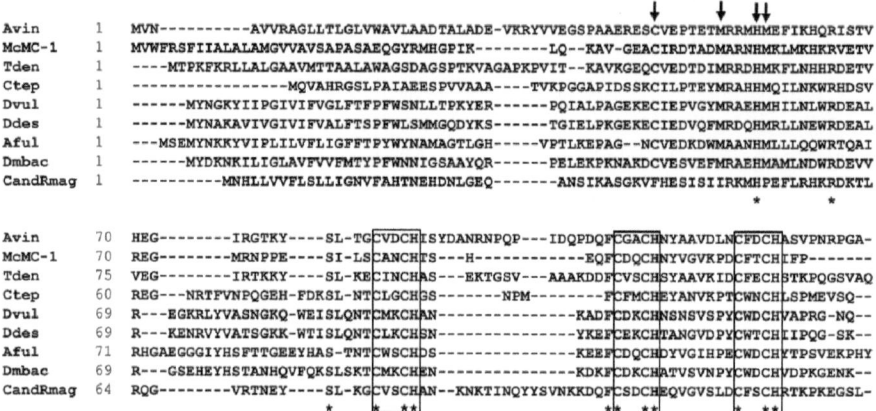

Figure 1.2: Sequence alignment of DsrJ proteins; amino acids 1-132 (*A. vinosum* numbering) shown only; heme binding motifs are boxed; putative distal heme ligands are indicated by arrows; strictly conserved residues are marked with an asterisk; *Allochromatium vinosum* (Avin), *Magnetococcus sp. MC-1* (McMC-1), *Thiobacillus denitrificans* (Tden), *Chlorobaculum tepidum* (Ctep), *Desulfovibrio vulgaris Hildenborough* (Dvul), *Desulfovibrio desulfuricans* ATCC 27774 (Ddes), *Archaeoglobus fulgidus* (Aful), *Desulfomicrobium baculatum* (Dmbac), *Candidatus* Ruthia magnifica (CandRmag).

DsrM

The sequence of *A. vinosum* DsrM can be found on the NCBI homepage using the accession number YP_003443228. Five transmembrane helices are predicted by all applied software. Using the conserved domain search tool, a domain of the nitrate reductase gamma subunit superfamily (cl00959) can be found in DsrM. Members of the same family are HmeC from *Archaeoglobus fulgidus*, the *E. coli* nitrate reductase (NarI) and the subunit E of the heterodisulfide reductase from *Methanosarcina* species. In particular NarI is a well characterized protein containing two hemes *b* (Magalon et al., 1997). The distribution of the transmembrane helices in DsrM and NarI is highly similar and the histidines, that ligate the hemes in NarI, are conserved in DsrM (Pott & Dahl, 1998). For NarI it has been shown that one heme is oriented on the periplasmic side of the membrane while the second heme is located on the cytoplasmic side of the membrane (Magalon et al., 1997). The calculated molecular weight of DsrM is 27.9 kDa and the theoretical isoelectric point is 9.85 (Pott & Dahl, 1998).

1 Introduction

DsrP

Based upon sequence analysis, *A. vinosum* DsrP (accession number: YP_003443233) is another membrane protein and ten transmembrane helices are predicted by several programs. It belongs to the NrfD protein family. Members of that family provide the membrane-spanning, quinone-interacting subunits to a number of complex iron–sulfur molybdoenzymes (CISM) (Rothery et al., 2008). DsrP shares highest sequence similarity with the PsrC subunit of polysulfide reductases from several organisms. DsrP is furthermore related to HybB, the membrane subunit of the *E. coli* hydrogenase 2 (Menon et al., 1994). DsrP has a calculated molecular weight of 45.6 kDa and an isoelectric point of 7.84. When the DsrMKJOP complex was purified from *D. desulfuricans* and from *Archaeoglobus fulgidus* (as Hme complex) DsrP was not detected in SDS-PAGE most probably due to the highly hydrophobic character (Pires et al., 2006; Mander et al., 2002).

DsrK

The analysis of the amino acid sequence of *A. vinosum* DsrK (accession number YP_003443229) reveals no putative signal peptide for the translocation into the periplasm and the protein is therefore thought to reside in the cytoplasm. The protein is related to HmeD from *Archaeoglobus fulgidus*, Hmc6 from *D. vulgaris* and to the D subunit of the heterodisulfide reductase from *Methanosarcina barkeri*. Based upon sequence analysis, DsrK contains two classical $CX_2CX_2CX_3C$ binding sites for [4Fe-4S] clusters that are also found in HdrD from *M. barkeri* (Künkel et al., 1997). The latter protein moreover contains two copies of a domain that has been designated as CCG domain in the Pfam database (accession number PF02754). In this domain up to five cysteines with the sequence $CX_nCCGX_mCX_2C$ are usually found where a tandem cysteine (CC) is usually followed by a glycine that led to the designation of the CCG domain. In DsrK the C-terminal motif is fully conserved while only the last cysteine of the N-terminal domain is conserved (Pires et al., 2006). It has been found that the C-terminal CCG domain – the one that is conserved in DsrK - binds another [4Fe-4S] cluster in HdrD and SdhC (Hamann et al., 2007; Hamann et al., 2009), which is therefore also postulated for DsrK (Pires et al., 2006). In some DsrK proteins the last cysteine of the CCG domain is replaced by an aspartate (Pires et al., 2006). However, aspartate has also been reported to be a ligand to FeS clusters (Lemos et al., 2002). Furthermore only four cysteines are required for the binding of a [4Fe-4S] cluster. The calculated molecular mass of DsrK is 58.5 kDa and its isoelectric point is calculated to be 4.7 (Pott & Dahl, 1998).

DsrO

A. vinosum DsrO (accession number: YP_003443232) is a ferredoxin-like protein and is related to the electron-transfer subunits of several CISM including tetrathionate reductase (TtrB), polysulfide reductase (PsrB) and nitrite reductase (NrfC). These proteins are associated with integral membrane subunits and transfer electrons from this subunit to a catalytic subunit (Rothery et al., 2008). Another electron-transfer subunit of a CISM is the hydrogenase 2 from *E. coli* (HybA) that accomplishes a contrary electron flow (Dubini et al., 2002). DsrO contains a typical signal peptide for translocation across the membrane via the Tat pathway (Berks et al., 2005) and is therefore thought to reside in the periplasm. In *A. vinosum* DsrO, four [4Fe-4S] clusters are predicted by sequence analysis (Dahl et al., 2005), while only three FeS cluster binding sites are predicted for the proteins from several *Desulfovibrio* species (Pires et al., 2006). No transmembrane helix is predicted to be present in DsrO in contrast to the respective subunits of the related Hmc or 9Hc complexes from *D. vulgaris* (Keon & Voordouw, 1996; Saraiva et al., 2001). The calculated molecular weight of DsrO is 28.9 kDa and its isoelectric point is 6.14.

1.5 Hemes and iron-sulfur clusters

Prosthetic groups are of essential importance for a transmembrane complex that is thought to participate in electron transfer. Prosthetic groups as hemes and FeS clusters contain iron and can transfer electrons due to the ability of the iron to change its valency from +3 to +2. Especially the hemes are in the focus of this work and their biochemical and biophysical properties were investigated. The prosthetic groups are therefore briefly introduced in this section.

Hemes

Hemes are prosthetic groups that consist of an iron atom located in the center of a protoporphyrin molecule. Basically four different types of hemes can be divided by their structure (heme *a*, *b*, *c* and *d*) although further structural variants exist. The structures of heme *b* and heme *c* are displayed in Figure 1.3. The structures are highly similar with the exception that the two vinyl side chains of heme *b* are replaced by covalent thioether bonds in heme *c*. These bonds involve the sulphydryl groups of cysteine residues provided by the apoprotein.

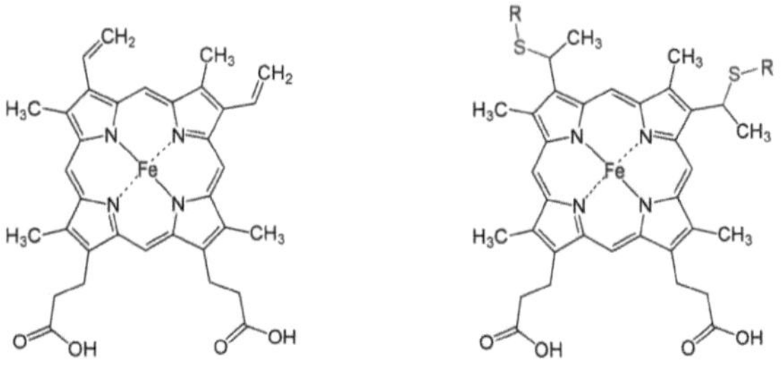

1 Introduction

Figure 1.3: Structure of heme *b* and heme *c*

A number of distinct protein classes bind hemes. These include catalases, peroxidases, globins and cytochromes. The latter are of central interest for the present work and will be considered in the following. Cytochromes have been defined as heme proteins whose principle biological function is the transport of electrons and/or hydrogen (Lemberg & Barrett, 1973). However, this definition is not very strict and accordingly some hemoproteins, such as P450 and the nitric oxide synthase, have also been termed "cytochromes". Although both proteins can be considered as *b*-type cytochromes, their main function is catalysis (Hasler et al., 1999; Alderton et al., 2001).

Cytochromes can either be classified by the heme type that is bound (see above) or by the heme (iron) ligation. The redox potential of cytochromes is usually in the range of -500 to +450 mV (Reedy et al., 2008) and is significantly influenced by the heme ligation (Paoli et al., 2002; Reedy et al., 2008; Battistuzzi et al., 2002; Wallace & Clark-Lewis, 1992). In the heme molecule the iron is ligated by the four nitrogen atoms provided by the protoporphyrin ring and is therefore placed in-plane of the molecule. Two ligand positions can be occupied additionally in the axial position. In cytochromes these ligands are usually amino acids provided by the protein.

In *c*-type cytochromes the heme is attached covalently to the protein. The proteins are characterized by a heme binding motif that has typically the sequence CXXCH. While the cysteines provide thioether linkages to the heme, the adjacent histidine serves as fifth ligand to the heme iron. This ligand is referred to as the proximal axial ligand. As an anomaly, the proximal heme ligand of one of the hemes in NrfA from several bacteria is a lysine (Simon, 2002). On the opposite site of the proximal ligand a sixth ligand (referred to as the distal axial ligand) can additionally ligate the iron leading to an octahedral complex (Figure 1.4). However, in many cases a sixth ligand is absent resulting in a five-coordinated high spin heme. In this heme the iron is slightly drawn outside of the plane of the porphyrin molecule. The same is observed when a weak ligand serves as sixth ligand. This results in a six-coordinated high spin heme.

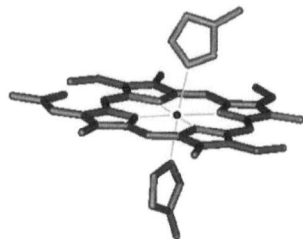

Figure 1.4: Schematic presentation of a six-coordinated His/His ligated heme; ligand bounds to the central iron atom are depicted in light grey.

The sixth ligand in c-type cytochromes is usually a histidine or a methionine. In particular in multiheme cytochromes c, the hemes usually exhibit His/His coordination (Pereira & Xavier, 2005). Nervertheless, there are amino acids, distinct from histidine or methionine, that can serve as axial ligands. His/Cys ligated hemes have for instance been observed in exceptional cases. His/Cys coordination in c-type cytochromes was first discovered in the *Rhodovulum sulfidophilum* SoxXA complex involved in thiosulfate oxidation (Cheesman et al., 2001) and has since also been studied in SoxXA from other organisms (Kappler et al., 2004; Ogawa et al., 2008; Rother & Friedrich, 2002). Besides this, His/Cys ligation was only proven to be present in one further c-type cytochrome. This is PufC, a protein that is associated with the photosynthetic reaction center in *Rhodovulum sulfidophilum* (Alric et al., 2004). Furthermore, the green heme protein from *Halochromatium salexigens* may also contain a His/Cys ligated heme (van Driessche et al., 2006). As already mentioned, upon sequence analysis a His/Cys ligated heme is expected to be present in DsrJ. EPR analysis of the purified DsrMKJOP complex from *Desulfovibrio desulfuricans* ATCC 27774 indeed revealed the presence of a His/Cys ligated heme (Pires et al., 2006) though experimental identification of the ligating cysteine was not conducted.

Iron-sulfur clusters

Iron-sulfur clusters fulfill multiple purposes in all life forms (Beinert et al., 1997). They can have a function in electron transfer, act as catalytic clusters or function as sensors of oxygen or iron. The most common forms are Fe_2S_2, Fe_3S_4, and Fe_4S_4 clusters (Figure 1.5).

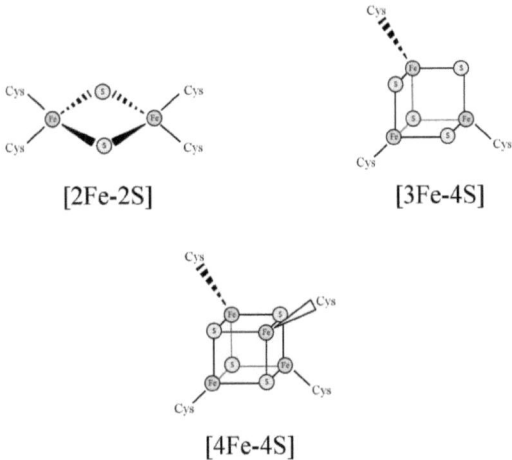

Figure 1.5: Most common forms of FeS clusters

The clusters are usually coordinated inside of the protein via three or four cysteine residues while in the Rieske iron sulfur proteins two cysteines and two histidines serve as ligands (Rieske et al., 1964). The redox potential of iron sulfur clusters is determined by the protein structure and usually lies between -650 mV and +450 mV (Beinert, 2000). Examples for FeS proteins are ferredoxins that usually have a low redox potential (-650 to -100 mV), rubredoxins having a moderate redox potential (-80 to +40 mV), or high potential iron sulfur proteins (HiPIP) that are known to have extremely positive redox potentials (up to +450 mV) (Beinert, 2000). Iron sulfur clusters are also combined with other cofactors such as in the siroheme-[4Fe-4S] in dSiR (compare section 1.3) and in the FeMoco of nitrogenase (Rubio & Ludden, 2008; Beinert, 2000).

1.6 Aim of this work

The aim of the present work was the characterization of the DsrMKJOP transmembrane complex to gain insights into its role in the oxidation of stored sulfur in *A. vinosum* with a special focus on the heme containing proteins. In particular the characterization of the unusual *c*-type cytochrome DsrJ was the primary task due to the predicted unusual His/Cys ligation of one of the hemes. A study of the individual components permits a more straightforward analysis of the redox cofactors, without interference from the other components as it occurs in the whole complex. Therefore, the major strategy was the individual production of the recombinant proteins in *E. coli* followed by their purification and biochemical and biophysical characterization. Another task was the enrichment of DsrMKJOP from *A. vinosum* to characterize the entire complex initially and to verify the results obtained for the recombinant proteins.

2 METHODS

2.1 Bacterial strains, antibodies, plasmids and oligonucleotides

Table 2.1: Bacterial strains used in this study

Bacterial strains	Genotype or phenotype	Source or reference
Escherichia coli strains		
DH5α	F⁻ Φ80d*lacZ*ΔM15 Δ(*lacZYA-argF*)*U169 recA1 endA1 hsdR17* (r_K^- m_K^+) *supE44* λ⁻ *thi-1 gyrA relA1*	Hanahan (1983)
S17-1	294 *(recA pro res mod⁺)* Tp^r Sm^r (pRP4-2-Tc::Mu-Km::Tn7)	Simon et al. (1983)
BL21 (DE3)	F⁻ *ompT hsdS$_B$* (r_B^- m_B^-) *gal dcm met* (DE3)	Novagen
C41 (DE3)	derived from BL21 (DE3) at least one uncharacterized mutation	Miroux & Walker (1996)
C43 (DE3)	derived from C41 (DE3) at least two uncharacterized mutations	Miroux & Walker (1996)
Allochromatium vinosum strains		
Rif50	Rif^r, spontaneous rifampicin-resistant mutant of *A. vinosum* DSMZ 180^T	Lübbe et al. (2006)
Δ*dsrM*	Rif^r, *in frame* deletion in *dsrM*	Sander et al. (2006)
Δ*dsrK*	Rif^r, *in frame* deletion in *dsrK*	Sander et al. (2006)
Δ*dsrJ*	Rif^r, *in frame* deletion in *dsrJ*	Sander et al. (2006)
Δ*dsrO*	Rif^r, *in frame* deletion in *dsrO*	Sander et al. (2006)
Δ*dsrP*	Rif^r, *in frame* deletion in *dsrP*	Sander et al. (2006)

Table 2.2: Primary antibodies used in this study

Antibody	Raised against peptide / application	Source or reference
Anti-DsrJ	H$_2$N-DANRNPQPIDQPDQFC-CONH$_2$	Sander (2005)
Anti-DsrM	H$_2$N-SPTRNQVDNPREQRHI-CONH$_2$	Sander (2005)
Anti-DsrK	H$_2$N-CDLDDPNEEEETDEAA-CONH$_2$	Dahl et al. (2005)
Anti-DsrO	H$_2$N-SQRLREIPSRQIREDL-CONH$_2$	Sander (2005)
Anti-DsrC	raised against recombinant DsrC	Dahl et al. (2005)
Anti-His	His-tag detection	Novagen/Merck (Darmstadt, Germany)
Anti-Strep	Strep-tag detection	IBABioTagnologies, (Göttingen, Germany)

2 Methods

Table 2.3: Plasmids used in this study

Plasmid	Genotype or description	Source or reference
pET11a	Ap^r, T7lac promoter	Novagen/Merck (Darmstadt, Germany)
pET22b	Ap^r, His-tag, T7lac promoter:	Novagen/Merck (Darmstadt, Germany)
pPR-IBA-1	Ap^r, Strep-tag, T7 promoter	IBABioTagnologies (Göttingen, Germany)
pASK-IBA-3	Ap^r, Strep-tag, Tet promoter	IBABioTagnologies (Göttingen, Germany)
pEC86	Cm^r, product of pEC66 and pACYC184 with the *E. coli ccmABCDEFGH* genes	Arslan et al. (1998)
pK18*mobsac*B	Km^r, *lacZ`*, *sacB*, Mob$^+$	Schäfer et al. (1994)
pACYCisc	Cm^r, Tet^r, Expression of the *isc* operon from *E. coli*	Gräwert et al. (2004)
pETJHis	Ap^r, PelB-leader, His-tag, *NcoI-XhoI* fragment of PCR-amplified dsrJ in pET22b	Schneider (2007)
pETJHisSP	Ap^r, native signalpeptide, His-tag, *NdeI-XhoI* fragment of PCR-amplified *dsrJ* in pET22b	This work
pPR-IBApelB	Ap^r, PelB-leader, Strep-tag *NcoI-XbaI* fragment from pET22b in *NcoI-XbaI* digested pPR-IBA1	This work
pPRIBApelBJStrep	Ap^r, PelB-leader, Strep-tag, *NcoI-Eco*47III fragment of PCR-amplified *dsrJ* in pPR-IBApelB	This work
pETJStrep	Ap^r, PelB-leader, Strep-tag, *NcoI-Hin*dIII fragment pPR-IBApelB in digested pET22b	This work
pETJC46SStrep	Ap^r, C46S mutation introduced into pETJStrep	This work
pBBR1MCS2-L	Km^r, *dsr* promoter and *dsrL* in pBBR1MCS2	Lübbe et al (2006)
pBBRJ	Km^r, *NdeI-Bam*HI fragment of PCR amplified *dsrJ* in digested pBBR1MCS2-L	This work
pBBRJC46S	Km^r, C46S mutation introduced into pBBRJ	This work
pBBRJHis	Km^r, His-tag coding sequence fused to *dsJ* in pBBRJ	This work
pMexN	Ap^r, His-tag, *NdeI-Bam*HI fragment of amplified *dsrM* in pET15b	This work
pMexC	Ap^r, His-tag, *NdeI-Bam*HI fragment of amplified *dsrM* in pET11a	This work
pPexC	Ap^r, His-tag, *NdeI-Eco*RI fragment of amplified *dsrP* in pET22b	This work
IBAdsrK	Ap^r, Strep-tag, *Eco*RI-*Eco*47III fragment of amplified *dsrK* in pPRIBA-1	This work
IBAdsrO	Ap^r, Strep-tag, *Eco*RI-*XhoI* fragment of amplified *dsrO* in pASKIBA-3	This work
pETCEX	Ap^r, His-tag, *NdeI-Bam*HI fragment of amplified *dsrC* in pET15b	Cort et al. (2008)

2 Methods

Table 2.4: Oligonucleotides used in this study, restriction sites are printed in bold.

Oligo-nucleotide	Sequence (5' → 3')	Source or reference
Expression of *dsrJ* in *E. coli*		
LdsrJNco-f	GCCC**CCATGG**ATGAGGTCAAGC	Schneider (2007)
LdsrJXho-r	CTTCAT**CTCGAG**GTGCCCTCCC	Schneider (2007)
fwdsrJanker	CTTGGAGGCG**CATATG**GTCAACG	This work
DsrJEco47III-r	CTTCAT**AGCGCT**GTGCCCTCCC	This work
Cys46Ser exchange		
DsrJC46S-f	CGGGAGTCCAGTGTCGAA	This work
DsrJC46S-r	TTCGACACTGGACTCCCG	This work
T7-f	TAATACGACTCACTAATGG	This work
LJHisXbar	GCAGC**TCTAGA**TCAGTGGTGGTGG	Schneider (2007)
Expression of *dsrM* in *E. coli*		
M1f	GGTGCGGA**CATATG**GCGTTTCTGAC	Sander (2005)
M1r	CGCTCGAATCA**GGATCC**TCATGACTGC	Sander (2005)
Mhis2rev	GGAGG**GGATCC**TCAGTGATGATGATGATGATGTGACTGCTCGAGC	This work
Expression of *dsrP* in *E. coli*		
DsrPNde-f2	TATC**CATATG**AAACGAGTCGTCTATC	This work
DsrPEco-r	CATCT**GAATTC**GGCTTGGCG	This work
Expression of *dsrK* in *E. coli*		
DsrKEcoRI-f	GAGGT**GAATTC**ATGGCCAAGGC	This work
DsrKEco47III-r	GGTCA**AGCGCT**CTCGTCTGTTTC	This work
Expression of *dsrO* in *E. coli*		
DsrOEcoRI-f2	CACAG**GAATTC**GGCCGCG	This work
DsrOXho-r	CTCG**CTCGAG**AAGGCTGAA	This work
Construction of *A. vinosum* Mhis2		
dsrMHis-forwA	CTCGCTGG**AAGCTT**GCGGACTGACCCAGGA	This work
dsrMHis-revA	ATGATGGTGATGGTGATGCATGGTACCGCACCCTCGGTGGAG	This work
dsrMHis-forwB	GGTACCATGCATCACCATCACCATCATGCGTTTCTGACGAACT	This work
dsrMHis-revB	ACCGC**TCTAGA**CGCAGGCGTCCATGT	This work
Complementation of *A. vinosum* Δ*dsrJ*		
JBam-r	GTTGT**GGATCC**GTTCATGGTTC	This work
JHisBam-r	GCAGC**GGATCC**TCAGTGGTG	This work
PCR probe for Southern blot analysis		
cvseq16F	AACTGGTACGGAGAAAGA	Pott-Sperling (2000)
PO-SK 2	CGACTTTCAAACCGATGG	Pott-Sperling (2000)

2.2 Chemicals, enzymes and kits

2.2.1 Chemicals

4-chloro-1-naphthol	Sigma (Taufkirchen, Germany)
anhydrotetracycline	IBA BioTagnologies (Göttingen, Germany)
anti-digoxigenin-AP	Roche (Mannheim, Germany)
blocking reagent	Roche (Mannheim, Germany)
CDP-Star	Roche (Mannheim, Germany)
dig-dUTP	Roche (Mannheim, Germany)
D-desthiobiotin	IBA BioTagnologies (Göttingen, Germany)
n-dodecyl-β-D-maltoside	Glycon (Luckenwalde, Germany)
monobromobimane	Fluka (Taufkirchen, Germany)
menadione	Sigma (Taufkirchen, Germany)
rifampicine	Fluka (Taufkirchen, Germany)
SDS	Serva (Heidelberg, Germany)

All other chemicals were obtained from the companies: Fluka (Taufkirchen, Germany), Merck (Darmstadt, Germany), Roth (Karlsruhe, Germany) and Sigma-Aldrich (Steinheim, Germany). All chemicals were at least of p. a. quality.

2.2.2 Enzymes

CIAP	Fermentas (St. Leon-Rot, Germany)
desoxyribonuclease II	Roth (Karlsruhe, Germany)
lysozyme	Sigma-Aldrich (Munich, Germany)
restriction endonucleases	Fermentas (St. Leon-Rot, Germany)
ribonuclease A	Roth (Karlsruhe, Germany)
T4-ligase	Fermentas (St. Leon-Rot, Germany)

2.2.3 Kits

BCA Protein Assay	Pierce (Rockford, USA)
GeneJET Plasmid Miniprep Kit	Fermentas (St. Leon-Rot, Germany)
GeneJET Gel Extraction Kit	Fermentas (St. Leon-Rot, Germany)

2.3 Software and online tools

AMPHIASEEK	prediction of amphiphatic helices
	http://npsa-pbil.ibcp.fr/cgi-bin/npsa_automat.pl?page=/NPSA/npsa_amphipaseek.html
Bioedit	sequence alignment editor
	available for download at http://www.mbio.ncsu.edu/BioEdit/bioedit.html
BLAST	nucleotide or protein sequences database search
	http://blast.ncbi.nlm.nih.gov/Blast.cgi
Clone Manager 9	sequence processing
	Sci-Ed software (Cary, USA)
Clustal W	sequence alignment
	http://www.ebi.ac.uk/Tools/clustalw2/index.html
Expasy	link collection for various DNA and protein analysis tools
	http://www.expasy.ch/tools/
LipoP	prediction of signal peptides
	http://www.cbs.dtu.dk/services/LipoP/
MEMSAT-SVM	prediction of α helices
	http://bioinf.cs.ucl.ac.uk/psipred/
MoluCAD	modeling and visualization tool for molecules
	New River Kinematics (Williamsburg, USA)
Office 2007	text and data processing
	Microsoft (Redmond, USA)
Olis SpectralWorks	control of the Olis DW-2
	On-Line Instrument Systems, Inc.
Origin Pro8	graph processing
	OriginLab corporation (Northampton, USA)
Photoshop Elements	image processing
	Adobe Systems (San Jose, USA)
SignalP	prediction of signal peptides
	http://www.cbs.dtu.dk/services/SignalP/
UV WinLab	control of the Perkin Elmer Lambda 11
	Perkin Elmer Inc. (Waltham, USA)
wheelAPP	helical wheel projection applet
	http://cti.itc.virginia.edu/~cmg/Demo/wheel/wheelApp.html
WinAspect	control of the analytic Jena Specord 210
	Analytic Jena AG (Jena, Germany)

2.4 Other materials

Standards

1 kb DNA ladder	Invitrogen (Karlsruhe, Germany)
100 bp DNA ladder	Invitrogen (Karlsruhe, Germany)
PageRuler Prestained Protein Ladder	Fermentas (St. Leon-Rot, Germany)
PageRuler Unstained Protein Ladder	Fermentas (St. Leon-Rot, Germany)

Chromatography material

Ni-NTA agarose	Quiagen (Hilden, Germany)
Strep-Tactin superflow	IBA BioTagnology (Göttingen, Germany)
Superdex 75 pg	Pharmacia (Uppsala, Sweden)

Other

3MM Chromatography paper	Whatman (Maidstone, United Kingdom)
Anaerocult A, Anaerotest	Merck (Darmstadt, Germany)
Centriplus Centrifugal Filter Device	Millipore (Schwalbach, Germany)
developer	Kodak (Rochester, USA)
dialysis tube	Serva (Heidelberg, Germany)
fixer	Kodak (Rochester, USA)
Immobilon-P (PVDF membrane)	Millipore (Schwalbach, Germany)
Protran BA 85 cellulose nitrate membrane	Schleicher & Schuell (Dassel, Germany)
Roti-Nylon plus membrane	Roth (Karlsruhe, Germany)
Vivaspin 500	Sartorius (Göttingen, Germany)
X-ray films X-OMAT AR	Kodak (Rochester, USA)

2.5 Microbiological methods

2.5.1 Media

E. coli was cultivated on LB-, 2×YT-, or NZYCM medium.

Luria Bertani (LB) medium (Maniatis et al., 1982)

tryptone	10 g
yeast extract	5 g
NaCl	5 g
dH_2O	ad 1000 ml
pH 7.5	

2 × YT medium (Maniatis et al., 1982)

tryptone	16 g
yeast extract	10 g
NaCl	5 g
dH_2O	ad 1000 ml
pH 7.0	

2 Methods

NZCYM medium (Blattner et al., 1977)

casein enzymatic hydrolysate	10 g
yeast extract	5 g
NaCl	5 g
casaminoacids	1 g
$MgSO_4 \times 7\ H_2O$	2 g
pH 7.5	

RCV medium (Weaver et al., 1975)

A. vinosum was cultivated on RVC medium when cells were grown for protein purification or western blot analysis.

RÄH medium	50 ml
yeast extract	0.5 g
NaOH	1.8 g
dH_2O	ad 1000 ml
pH 7.0	

RÄH medium

malate	60 g
NH_4Cl	24 g
$MgSO_4 \times 7\ H_2O$	4 g
$CaCl_2 \times 2\ H_2O$	1.4 g
SL12 (10×)	20 ml
dH_2O	ad 1000 ml

Trace element solution SL12 (Pfennig & Trüper, 1992)

EDTA	30 g
$FeSO_4 \times 7\ H_2O$	11 g
H_3BO_3	3 g
$CoCl_2 \times 6\ H_2O$	1.9 g
$MnCl_2 \times 4\ H_2O$	0.5 g
$ZnCl_2$	0.42 g
$NiCl_2 \times 6\ H_2O$	0.24 g
Na_2MoO_4	0.18 g
$CuCl_2 \times 2\ H_2O$	0.02 g
dH_2O	ad 1000 ml

RÄH medium and the trace element solution were stored without sterilization. Before use of the RCV medium, 5 % (v/v) of a 180 mM K_2HPO_4-KH_2PO_4 buffer (pH 7.0) was added. For the preparation of solid RCV plates, 1.5 % agar was added to RCV medium before autoclaving.

0 medium (Hensen et al., 2006)

For the characterization of the phenotype *A. vinosum* mutant strains were grown in Pfennig medium lacking sulfide referred to as 0 medium (Hensen et al., 2006). Sterilized sulfide was directly added into the fermenter when the experiment was started. 0 medium was prepared by combining three solutions that were sterilized separately. The following solutions were used to prepare 10 liter of 0 medium:

Solution 1: salt solution prepared in a 10 liter carboy:

KCl	3.3 g
$MgCl_2 \times 6\ H_2O$	3.3 g
$CaCl_2 \times 2\ H_2O$	4.3 g
NH_4Cl	3.3 g
SL12 (10×)	10 ml
dH_2O	ad 8000 ml

Solution 2: Phosphate solution

KH_2PO_4	3.3 g
dH_2O	ad 1000 ml

Solution 3: Carbonate solution

$NaHCO_3$	15 g
dH_2O	ad 1000 ml

After autoclaving, the three solutions were combined under a nitrogen atmosphere. The cloudy medium was bubbled with CO_2 resulting in the loss of cloudiness. At this point the pH of the solution was between 6.5 and 6.8. The medium was filled into 1l-flasks that were tightly closed and stored in the dark until used.

2 Methods

For the isolation of *A. vinosum* mutants after conjugation, RCV plates were prepared as follows:

RÄH medium	6.25 ml
yeast extract	0.62 g
NaOH	0.225 g
NaCl	0.625 g
Phytagel	1.25 g
dH_2O	ad 125 ml
pH 7.0	

feeding solution

HNaS × H_2O	3.1 g
$NaHCO_3$	5 g
dH_2O	ad 100 ml

After autoclaving, 5 % (v/v) buffer (180 mM K_2HPO_4-KH_2PO_4 pH 7.0) was added as well as 0.25 % (v/v) feeding solution, 0.2 % (v/v) sodium acetate and 0.2 % (v/v) thiosulfate solution.

2.5.2 Antibiotics

Table 2.5 lists the antibiotics that were used for the cultivation of *A. vinosum* and *E. coli* strains.

Table 2.5: Antibiotics used in this study

antibiotic	solvent	final concentration [µg ml^{-1}]
A. vinosum		
rifampicin	methanol	50
kanamycin	dH_2O	10
E. coli		
ampicillin	dH_2O	100
kanamycin	dH_2O	50
chloramphenicol	ethanol	25

2.5.3 Cultivation of *A. vinosum* and *E. coli*

A. vinosum was cultivated photoorganoheterotrophically on RCV medium for standard applications. The cultures were either grown in glass flasks (50 – 1000 ml volume) or on solidified plates at 30 °C in an illuminated incubator (Biotron, Hilden, Germany). In the latter case the plates were placed in an anaerobic jar (Merck, Darmstadt, Germany). An anaerobic medium was generated inside by using the Anaerocult A reagent (Merck, Darmstadt, Germany).

For the characterization of the phenotype of an *A. vinosum* strain, the cultures were grown photolithoautotrophically in a glass fermenter at 30 °C under anoxic conditions. The fermenter was placed on a magnetic stirrer and the culture was illuminated by two 60 W lamps. During the experiment the pH was measured with a pH electrode and adjusted to pH 7.0 ± 0.1 by the addition of 1 M HCl or 1 M Na_2CO_3.

250 ml of a photoheterotrophically grown stationary-phase culture was harvested (5900 × g, 10 min) and the cell material was used to inoculate 1 l of 0 medium in the fermenter. The experiment was started by the addition of approx. 2 mM sulfide from a sterile stock solution (1 M). During the experiments samples were taken in regular intervals to monitor the following parameters: protein concentration of the culture, optical density of the culture at 690 nm, sulfate concentration, concentration of elemental sulfur and concentration of thiols.

For standard applications, *E. coli* was cultivated on solid LB agar plates at 37 °C or in liquid LB medium with an agitation of 180 rpm. Amongst others the following factors were verified within the production of recombinant proteins to achieve optimal protein production:

- *E. coli* strain
- medium
- temperature
- point of induction
- strength of induction
- expression time

The conditions that were best suited for the production of the individual proteins are described in the results section. *E. coli* cells were cultivated as described by Cort et al. (2001) for the production of DsrC.

2.5.4 Conservation

For long term storage of *A. vinosum* strains, a respective culture of *A. vinosum* was cultivated photoorganoheterotrophically on RCV medium for approximately 5 days. 50 ml of this culture were harvested (2500 × g, 20 min) and resuspended in 5-10 ml of RCV medium. This preparation was mixed with an equal volume of 10 % (v/v) sterile DMSO solution and stored in liquid nitrogen in Nunc Cryo Tubes (Roskilde, Denmark).

For long term storage of *E. coli* strains, an aliquot of a culture that was grown over night on LB medium was mixed with an equal volume of sterile glycerol, frozen in liquid nitrogen and stored at -70 °C.

2.5.5 Preparation of competent *E. coli* cells

Solutions: 2 × YT medium
$CaCl_2$ / $MgCl_2$ solution (70 mM $CaCl_2$, 20 mM $MgCl_2$)

The chemically competent *E. coli* cells used for transformation, were prepared using the calcium chloride method (Dagert & Ehrlich, 1979). 70 ml of 2 × YT medium were inoculated with 700 µl of an overnight grown culture of the respective *E. coli* strain. The culture was cultivated at 37 °C and 180 rpm until an OD_{600} of 0.3 to 0.4 was reached. Then the culture was harvested by centrifugation (1900 x g, 4°C, 6 min) and the pellet was resuspended in 21 ml of $CaCl_2$ / $MgCl_2$ solution. The cells were incubated in this solution on ice for 30 min and pelleted again. After resuspending in 7 ml $CaCl_2$ / $MgCl_2$ solution and another incubation on ice for 45 min, the preparation was mixed with 1.750 ml sterile glycerol, aliquoted and stored at -70 °C.

2.6 Analytical methods

2.6.1 Determination of protein concentration

Protein concentrations in *A. vinosum* cultures were determined using the method of Bradford (1976). 1 ml cell culture was centrifuged (4 min, 16000 × g) and the pellet was stored at -20 °C until used. For the assay, the samples were thawed and resuspended in 1 ml NaOH (1 M). After incubation at 95 °C for 5 min, the samples were centrifuged again and the supernatant was subjected to the protein determination using the Bradford reagent (Sigma, Taufkirchen, Germany) according to the manufacturer's instruction. The calibration line was generated with $0 - 1.4$ mg ml^{-1} BSA.

Protein concentrations of cell fractions or of samples containing purified protein were determined using the BCA Protein Assay Kit according to the manufacturer's instructions (Thermo Fisher Scientific, Waltham, USA). A calibration line was generated with 0 – 500 µg ml^{-1} BSA.

2.6.2 Heme quantification

Heme quantification was carried out by pyridine hemochrome spectra (Berry & Trumpower, 1987). This method allows the simultaneous determination of hemes *a*, hemes *b* and hemes *c*. Pyridine serves as axial ligand to the hemes and the cytochromes therefore absorb at distinct wavelengths with a distinct absorption coefficient independently from their native axial heme ligands. The quantity of hemes in a mixture of different hemes (i.e. in the DsrMKJOP preparations from *A. vinosum*) was calculated by using the inverse matrix of extinction coefficients of pyridine hemochromes from reduced minus oxidized difference spectra at five wavelengths. When only one type of heme was present in the sample, calculations were made with reduced minus oxidized difference spectra at two wavelengths and the recommended extinction coefficients (ε) were used.

for heme *c*: $E_{550} - E_{535}$ ε: 23.97

for heme *b*: $E_{557} - E_{540}$ ε: 23.98

2.6.3 Quantification of iron sulfur clusters

Iron sulfur clusters were quantified by the separate determination of non-heme iron and acid labile sulfur.

Determination of non-heme iron

Solutions: Trichloroacetic acid 20 % (w/v)
1,10-O-Phenanthrolin 0.1 % (w/v)
Ascorbic acid 60 mM
Ammoniumacetate solution (saturated)

A sample of 750 µl was mixed with 250 µl TCA and after 10 min incubation the sample was centrifuged (16000 × g, 5 min). 400 µl of the supernatant was diluted with 360 µl dH$_2$O and 150 µl of the Phenantrolin solution, 50 µl of the ascorbic acid solution and 40 µl of ammoiumacetate solution was added. After mixing of the sample, the extinction at 510 nm was measured referenced to a reagent blank. The calibration line was generated with 0 – 150 nm Fe(II)SO$_4$.

2 Methods

Determination of acid soluble sulfur

Solution A: (5 volumes of a 2.6 % (w/v) zincacetate solution
+ 1 volume of 6 % (w/v) NaOH in ddH$_2$O)
Solution B: 20 mM N`N`-Dimethyl-p-Phenylen-Diaminsulfate in 7,2 M HCl
Solution C: 30 mM FeCl$_3$ in 11,2 M HCl

A sample of 700 µl was mixed with 500 µl freshly prepared solution A and incubated for 60 min in the dark. 100 µl solution B were added as well as 100 µl solution C. After mixing and further incubation of 20 min, the extinction of the sample at 670 nm was determined referenced to a reagent blank. For the calibration line, 0 - 20 nmol sodium sulfide was used.

2.6.4 Determination of elemental sulfur

Solution: Ferric nitrate reagent 30 g Fe(NO$_3$)$_3$ × 9 H$_2$O, 40 ml 55 % HNO$_3$, ad 100 ml dH$_2$O

A cell pellet with up to 200 nmol sulfur was resuspended in 200 µl dH$_2$O and 100 µl of a sodium cyanide solution (0.2 M) were added. The preparation was incubated for 10 min at 100°C followed by the addition of 650 µl dH$_2$O and 50 µl ferric nitrate reagent. The preparation was centrifuged (16000 × g, 2 min) and the extinction at 460 nm was measured against a reagent blank. A calibration line was recorded with sodium thiocyanate (rhodanide) in a range of 0 to 300 nmol per assay.

2.6.5 Determination of thiols by HPLC

Solutions for derivatisation: HEPES buffer (50 mM HEPES, 5 mM EDTA, pH 8.0), monobromobimane solution (96 mM monobromobimane in acetonitrile), 65 mM methanesulfonic acid

Thiole compounds like sulfide, polysulfides, thiosulfate and sulfite can be quantified after derivatisation with the fluorescent dye monobromobimane (Rethmeier et al., 1997). Therefore, a 50 µl sample taken from the fermenter was mixed with 50 µl HEPES buffer and 55 µl acetonitrile. After the addition of 5 µl monobromobimane solution the preparation was mixed and incubated in the dark. After 30 min the reaction was stopped by the addition of 100 µl methanesulfonic acid, and the preparation was stored at -20°C for further use.

For the HPLC analysis, the samples were thawed and centrifuged at 16000 × g for 10 min. 10 µl of the supernatant were mixed with 190 µl of ddH$_2$O of which 50 µl were injected onto the column.

The conditions for the HPLC analysis were the following:

Solution A: 0.25 % (v/v) acetic acid, pH 4.0

Solution B: Methanol

Column: LiChrospher 100 RP18 125 – 4 (5 µm) (Merck, Darmstadt, Germany)

Elution protocol: Linear gradient

Table 2.6: Elution protocol for determination of thiols by HPLC

Time [min]	% solution A	% solution B
0	85	15
5	85	15
50	55	45
55	0	100
58	0	100
61	85	15
76	85	15

Flow rate: 1 ml min^{-1}

Temperature: 35 °C

Detection: Fluorescence detection, excitation at 380 nm, emission at 480 nm

The calibration line was recorded with 0 to 1 mM sodium sulfide.

2.6.6 Determination of sulfate by HPLC

For the Determination of sulfate a 1 ml sample taken from the fermenter was centrifuged (16000 × g, 4 min) and the supernatant was stored at -20 °C until further use. For the analysis the sample was thawed, centrifuged again and 100 µl of the resulting supernatant were injected onto the column. The conditions for the HPLC analysis were the following:

Solution A: 4 mM *p*-Hydroxybenzoic acid

Solution B: Methanol

Column: PRP-X100 (Hamilton, Bonaduz, Switzerland)

Elution protocol: isocratic; 97 % solution A, 3 % solution B

Flow rate: 2 ml min^{-1}

Detection: UV detection at 254 nm

The calibration line was recorded with 0 to 5 mM sodium sulfate

2.7 Molecular biological methods

2.7.1 Polymerase chain reaction (PCR)

For the amplification of DNA and introduction of restriction sites, standard PCR were carried out. The PCR were either run in a Trio-Block (Biometra) or in a MyCycler (Bio-Rad) thermal cycler. Amplification of DNA for subsequent cloning was carried out by the use of *Pfu* polymerase while *Taq* polymerase was used for control PCR. The PCR were run according to the manufacturer's instructions (Fermentas, St. Leon-Rot, Germany). The oligonucleotides were purchased from MWG. While purified chromosomal DNA or plasmid DNA were used when PCR were carried out for molecular cloning, colony PCR were used for the identification of positive mutants. In the latter case, cell material obtained from liquid culture or from a single colony on a plate was washed twice with ddH$_2$O and used as template. An initial denaturation step (95 °C, 10 min) was carried out prior to the addition of the polymerase.

2.7.2 GeneSOEING

Site directed mutagenesis was carried out by gene SOEing (gene splicing by overlap extension) developed by Horton (1995). This comprised the construction of the DsrJCys46Ser mutant and the fusion on the His-tag coding sequence to *dsrM* for the construction of the *A. vinosum* Mhis2 mutant. Briefly two DNA fragments were produced carrying the desired mutation at the end and at the beginning of the sequence, respectively. The two fragments served as template for a third PCR that results in a DNA fragment carrying the desired mutation within the gene.

2.7.3 Enzymatic DNA modification

Restriction enzyme digestion

Restriction endonucleases were purchased from Fermentas (St. Leon-Rot, Germany) and used according to the manufacturer's instructions. The respective buffer was chosen from the Fermentas five buffer system. Typically a restriction assay had a volume of 10 – 50 µl and restriction was carried out for 1-3 hours at the recommended temperature.

Alkaline phosphatase

For the construction of plasmids, the digested vector backbone was treated with the calf intestinal alkaline phosphatase purchased from Fermentas (St. Leon-Rot, Germany). In general 1 µl enzyme was added to the restriction assay and incubated for 30 min at 37 °C followed by heat inactivation at for 5 min.

Ligation

To insert DNA fragments into a plasmid, the restricted fragments were ligated using the T4 ligase (Fermentas, St. Leon-Rot, Germany). Ligations were performed either over night at 6 °C or for 1 h at 37 °C. Prior to transformation, the ligation assay was heat inactivated at 70 °C for 5 min.

2.7.4 DNA visualization

TAE: 40 mM Tris, 20 mM acetic acid, 10 mM EDTA
10 × loading buffer: 0.25 % (w/v) bromphenole blue, 40 % (w/v) sucrose

For electrophoretic DNA separation, the samples were mixed with the respective amount of 10 × loading buffer and applied to agarose gels (1-1.5 % (w/v) agarose in TAE). As a standard, 5 µL DNA ladder (Invitrogen, Karlsruhe, Germany) were applied. The gels were covered with TAE and run at 100 V. Gels were stained in ethidium bromide solution (10 µg ml^{-1}) for 10 min under gentle shaking in the dark and rinsed with dH$_2$O. The DNA bands were visualized on a UV transilluminator (INTAS, Göttingen, Germany).

2.7.5 DNA preparation and purification

Preparation of plasmid DNA

Buffer 1: 50 mM Tris, 10 mM EDTA, 100 µg ml^{-1} Ribonuclease A, pH 8.0
Buffer 2: 200 mM NaOH, 1 % (w/v) SDS, pH 12.5
Buffer 3: 3 M Potassium acetate, pH 5.5

Plasmid DNA was prepared from *E. coli* DH 5α either by using the GeneJet Plasmid MiniprepKit (Fermentas, St. Leon-Rot, Germany) or by the non column prep. For the latter, 1.5 ml *E.coli* culture were harvested by centrifugation (16000 × g, 3 min) and the pellet was resuspended in 200 µl buffer 1. 200 µl buffer 2 were added and the vials were inverted several times followed by the addition of 200 µl buffer 3 and mixing. The preparation was centrifuged (16000 × g, 3 min) and the supernatant was transferred into vials containing 500 µl chloroform. The preparation was mixed, centrifuged (16000 × g, 3 min), and the supernatant mixed with 500 µl isopropanol. After another centrifugation (16000 × g, 3 min) and discarding of the supernatant the pellet was dried and resuspended in 50 µl ddH$_2$O.

Purification of DNA from agarose gels

For cloning, DNA fragments were excised from the agarose gel and purified using the GeneJET Gel Extraction Kit (Fermentas, St. Leon-Rot, Germany), according to the manufacturer's instructions.

Preparation of chromosomal DNA from *A. vinosum*

TES buffer (100 mM NaCl, 10 mM TrisHCl, 1mM EDTA, pH 8.0
TE buffer (10 mM TrisHCl, 1mM EDTA, pH 8)
Saccharose solution (20 % (w/v) saccharose in TES)
Lysozyme RNAse solution (20 mg ml^{-1} lysozyme, 1mg ml^{-1} RNAse)
sarcosine solution (10 % (w/v) laurylsarcosine, 250 mM EDTA)

DNA was isolated from *A. vinosum* strains by sarcosyl lysis (Bazaral & Helinski, 1968). The bacteria were grown on RCV medium and harvested by centrifugation (4000 × g, 10 min). The pellet was washed in TE buffer and 50 mg of the cell material were resuspended in 2 ml ice-cold TES buffer. After centrifugation (16000 × g, 4°C, 10 min) 250 µl of saccharose solution were added to the pellet, and the preparation was incubated on ice for 30 min. 250 µl of lysozyme RNAse solution were added, followed by an incubation at 37 °C for 30 min with gentle shaking. 100 µl sarcosine solution were added and the sample was pressed through a sterile cannula (1.2 × 49 mm) to achieve shearing of the DNA that was subsequently purified via phenol / chloroform extraction. Therefore, the sample was mixed with an equal volume of phenol / chloroform / isoamylalcohol (25:24:1) and the DNA was extracted by powerful shaking. After centrifugation (16000 × g, 5 min), the supernatant was subjected to further purification steps. The final purification step was performed with chloroform / isoamylalcohol (24:1). The purified DNA was transferred into TE buffer by dialysis and stored at 4 °C. To determine the concentration and purity of the isolated DNA, absorption of a diluted sample was measured at 260 nm and 280 nm, corresponding to the absorption maxima of DNA and protein at the respective wavelengths.

2.7.6 DNA sequencing

The correct sequences of all inserts of the plasmids constructed in this study were verified by DNA sequencing. This was carried out by either Sequiserve (Vaterstetten, Germany) or GATC (Konstanz, Germany).

2.7.7 DNA transfer

Transformation

Foreign plasmid DNA was inserted into *E. coli* strains by transforming chemically competent cells (see section 2.5.5). 1 µl plasmid DNA or 5 – 10 µl of a ligation assay were added to 100 µl of cells. The preparation was first incubated for 30 – 120 min on ice followed by 90 s incubation at 42 °C. The assay was allowed to cool down for two min and 500 µl of 2 × YT medium were added and the mixture incubated at 37 °C for 45 – 60 min. Finally, different aliquots of the assay were applied onto LB plates containing the respective antibiotic. To insert two plasmids into a single strain, the whole procedure was done twice. In between the cells were prepared as described in section 2.5.5.

Conjugation

Insertion of plasmids into *A. vinosum* strains was carried out by conjugation with *E. coli* S17-1 carrying the respective plasmid by the method of Pattaragulwanit & Dahl (1995). Briefly, *A. vinosum* and *E. coli* cells were mixed in a ratio of 3 to 1 in equal volumes. The preparation was centrifuged (5000 × g, 5 min), resuspended in a little volume and applied onto a sterile cellulose nitrate filter (0.45 µm, Sartorius, Göttingen, Germany), which was lying on RCV solid medium. After two days of incubation under anoxic conditions in the light, the filter was placed inside of a sterile micro reaction tube and the cells were washed off with RCV medium. Cells were plated on RCV solid medium containing the appropriate antibiotics for the selection of transconjugants.

Plasmids, that were inserted into *A. vinosum* strains were either based on the plasmid pBBR1MCS2 (Kovach et al., 1995) or based on the plasmid pK18mobsacB (Schäfer et al., 1994). While the first is replicable in *A. vinosum* and thus allowing a plasmid based gene expression, the second is not replicable. Therefore, the plasmid is integrated into the genome of *A. vinosum* during the conjugation process. In a second step, the vector backbone is removed from the genome when the bacteria are cultivated for several days in growth medium, where the respective antibiotic is omitted. The cells can be selected by making use of the exoenzyme levansucrase that is encoded in the *sacB* gene in the vector backbone, since the expression of the *sacB* gene is lethal on a medium containing sucrose.

2.7.8 Southern blotting

20 × SSC	3 M NaCl, 0.3 M sodium citrate, pH 7.0
blocking solution 1	2 % (v/v) buffer A, 2.5 % (v/v) 20 × SSC, 300 µM N-lauroyl sarcosine, 60 µM SDS, 2 % (w/v) blocking reagent
blocking buffer 2	1 % (w/v) blocking reagent in buffer A
washing buffer	0.3 % (v/v) Tween 20 in buffer A
buffer A	100 mM maleic acid, 150 mM NaCl, pH 7.5
buffer TS	100 mM Tris, 100 mM NaCl, pH 9.5

Southern blots can be used for the detection of specific sequences among DNA fragments that were separated by gel electrophoresis (Southern) and were therefore applied for the identification of chromosomal mutations. Briefly, chromosomal DNA isolated from *A. vinosum* strains was digested with appropriate restriction enzymes and electrophoretically separated in an agarose gel. The gel was stained with ethidium bromide and documented (see section 2.7.4). Afterwards, the DNA was transferred onto a Roti-Nylon plus membrane (Roth, Karlsruhe, Germany) by a capillary blot and the membrane was shortly rinsed in 2 × SSC. After the DNA was covalently linked to the membrane by UV crosslinking (UV Stratalinker 1800, Stratagene), the membrane was used for hybridization.

Therefore, a DNA probe was generated by PCR that was suited for the detection of the desired DNA fragment. In the PCR assay, the nucleotide dTTP was replaced by digoxigenin-dUTP and the resulting DNA fragment was excised from the agarose gel and used for hybridization.

Prior to hybridization, the membrane was incubated in blocking solution 1 to prevent unspecific binding of the DNA probe. Blocking was performed in a Hybaid Mini hybridization oven under constant rotating for at least 4 hours at 68 °C. For hybridization, the DNA probe was added to 20 ml of blocking solution 1, incubated at 100 °C for 20 min, and poured over the membrane. After incubation over night in the hybridization oven the membrane was incubated twice in 2 x SSC + 1 % SDS solution for 5 min at room temperature. After 5 min incubation with washing buffer, blocking buffer 2 was applied to the membrane to saturate the unspecific binding sites for the antibody. After 5 min, the buffer was replaced by fresh blocking buffer containing 0.013 % (v/v) anti-digoxigenin alkaline phosphatase antibody conjugate. Unspecifically bound antibody was removed from the membrane by washing it in washing buffer for 15 min twice. Then the membrane was first incubated in buffer TS for 5 min followed by 20 min incubation in buffer TS to which the substrate for the alkaline phosphatase has been added (0.1 % (v/v) CDP-Star). The resulting emittance of light was detected on an X-ray film.

2.8 Working under anoxic conditions

Redox indicator: 4 g Tris, 0.2 g glucose, 1 mg resazurin, dH$_2$O ad 20 ml
H$_2$S absorber: 2.5 g Ag$_2$SO$_4$, 5 mM H$_2$SO$_4$, 500 ml glycerol, ad 1000 ml dH$_2$O

Experiments that were known or thought to be hindered by the presence of oxygen were carried out in an anaerobic chamber (Coy, Grass Lake, USA). The chamber provides a strict anaerobic atmosphere of 0-5 ppm oxygen by the use of a palladium catalyst and hydrogen gas mix. The chamber was filled with 98 % N$_2$ and 2 % H$_2$. The absence of oxygen was controlled with a combined oxygen and hydrogen analyzer (Coy). Additionally a chemical redox indicator was brought inside of the camber that was prepared under oxic conditions, boiled until it was colorless and then placed inside of the chamber. For the reconstitution of FeS clusters, dishes filled with H$_2$S absorber were placed inside of the chamber to prevent the sulfide from damaging the palladium catalysts. All liquids used were degassed and brought inside of the chamber at least 18 hours before the experiments were started.

2.9 Proteinbiochemical methods

2.9.1 Cell lysis

Lysis by sonication
Small amounts of cells were lysed by sonication. Therefore, the cells were resuspended in 5 ml lysisbuffer per g wet weight and the suspension was subjected to ultrasonic treatment (Cell Disruptor B15, Branson) at 50 % capacity for 1.5 min per ml. Some grains of desoxyribonuclease II were added to the lysate and incubated on ice for 15 min followed by a centrifugation step (25000 × g, 30 min, 4 °C). The supernatant that contained the soluble and membrane fractions was designated as raw extract, while the pellet contained the insoluble cell fraction.

Lysis by French press
When larger amount of cells were lysed or when the lysis was carried out under anoxic conditions, a French pressure cell press (Thermo Fisher Scientific, Waltham, USA) was used to disrupt the cells. The cells were resuspended in 2 - 3 ml lysisbuffer, and the preparation was passed through the French press at 16000 psi for at least two times. Afterwards the preparation was centrifuged as described for the lysis by sonication.

2.9.2 Ultracentrifugation

For the preparation of the membrane fraction, raw extracts were applied to ultracentrifugation (145 000 × g). In case of *E. coli* raw extracts, ultracentrifugation was carried out for 2 hours while this time was expanded to 3 hours, when *A. vinosum* raw extracts were used. The pellet after centrifugation contained the membrane fraction, whereas the supernatant contained the soluble fraction.

2.9.3 Solubilization of membrane proteins

Solubilization of membranes is obligatory for the purification of membrane proteins. Therefore, membranes were prepared by ultracentrifugation (2.9.2) and the resulting pellet was resuspended in lysis buffer. In order to solubilize the membranes, the detergents *n*-dodecyl-β-D-maltoside (DDM) or Triton X-100 were added from stock solutions. The preparation was gently stirred on ice for 2 – 3 hours or overnight. After subsequent ultracentrifugation, the solubilized proteins were found in the supernatant.

2.9.4 Chromatography methods

His-Tag purification
Lysis buffer: 50 mM NaH_2PO_4, 300 mM NaCl, pH 7.5

Recombinant proteins equipped with a hexahistidine-tag (His-tag) were purified via immobilized metal ion affinity chromatography (IMAC) using Ni-NTA agarose from Quiagen (Hilden, Germany). This method allows the use of DDM that is a powerful but mild detergent, widely used in the purification of membrane proteins and especially of membrane protein complexes. It was therefore chosen for purification of the membrane proteins that were heterologously produced in *E. coli* and also for the purification of His-tagged DsrJ from *A. vinosum*.
Binding of the proteins to the Ni-NTA agarose was performed in the batch mode. Therefore, the solubilized protein extract was mixed with the Ni-NTA agarose on ice. The mixture was incubated for 30 – 60 min with very gentle stirring or shaking. The mixture was the applied into polypropylene tubes and the column was washed with a step gradient of increasing imidazole concentration as recommended by the manufacturer and finally the proteins were eluted from the column. All buffers contained 0.1 % (w/v) DDM when membrane proteins were purified.

Strep-tag purification

Strep-tag purification is another affinity chromatography that is based on the high affinity of the Strep-tag to Strep-Tactin. It usually provides more efficient purification compared to the His-Tag system and was therefore used for the purification on DsrJ, DsrK and DsrO. Purification of DsrJ was switched from the His-tag to the Strep-tag to avoid a putative interaction of the six histidines with hemes that were ligated by weak ligands (i.e. methionine or cysteine) in the native enzyme. Furthermore imidazole is omitted in the elution buffer that can also ligate heme iron. Strep-Tactin superflow was purchased from IBA BioTagnology (Göttingen, Germany) and the purification was carried out according to the manufacturer's instructions.

Gel filtration

Gel filtration of recombinant DsrJ was accomplished by the help of a HiLoad system (Pharmacia, Uppsala, Sweden). The protein that eluted from the Strep-Tactin column was centrifuged (16000 × g, 5 min, 4 °C) and immediately applied to a Superdex 75 pg column (Pharmacia) that was equilibrated with 50 mM NaH_2PO_4, pH 7.0. After the protein has been loaded onto the column, the flow rate was set to 0.5 ml min^{-1} and fractions of 1 ml were collected.

2.9.5 Electrophoretic protein separation

For the electrophoretic separation of proteins, polyacrylamide gel electrophoresis (PAGE) was performed. Depending on the properties of the proteins that were applied to PAGE, different electrophoreses were performed. The Bio-Rad Mini Protean system was used in each case. The gels were either applied to Coomassie staining, Western blotting or in-gel heme staining.

SDS-PAGE

For the separation of the proteins DsrO and DsrJ SDS-PAGE was performed by the method of Laemmli (1970).

4 × Gel buffer A:	1.5 M Tris, 0.3 % (w/v) SDS, pH 8.8
4 × Gel buffer B:	0.5 M Tris, 0.4 % (w/v) SDS, pH 6.8
Running buffer:	3 g Tris, 14.4 g glycine, 1 g SDS, ad 1000 ml ddH_2O
Acrylamide/Bisacylamide solution:	Rotiphorese 30 (Roth, Karlsruhe, Germany)
APS solution:	10 % (w/v) APS in ddH_2O

The resolving gel was prepared by mixing appropriate volumes of Rotiphorese 30, gel buffer A and ddH_2O to obtain gels with 10 or 15 % (v/v) acrylamide/bisacrylamide. Then TEMED and APS solution were added and the gel was poured. After the resolving gel got solid, the stacking gel was poured accordingly with 4.5 % (v/v) acrylamide/bisacrylamide and the use of gel buffer B. Gels were run at 60 V until the proteins entered the resolving gel. Then the voltage was set to 100 V until the run was complete.

Tricine-SDS-PAGE

3 × Gel buffer:	3 M Tris, 1 M HCl, 0.3 % (w/v) SDS, pH 8.45
10 × Anode buffer:	1 M Tris, 0.225 M HCl, pH 8.9
10 × Kathode buffer:	1 M Tris, 1 M Tricine, 1 % (w/v) SDS, pH ~8.25
Acrylamide/Bisacylamide solution:	Rotiphorese 30 (Roth, Karlsruhe, Germany)
APS solution:	10 % (w/v) APS in ddH$_2$O

Tricine-SDS-PAGE is a method that is well suited for the separation of membrane proteins (von Jagow & Schägger, 1994) and was therefore used for the separation of DsrM, DsrK and DsrP. Especially transferring of hydrophobic proteins onto inert membranes is significantly enhanced by this method (Schägger, 2006). The buffer system consists of Tricine-Tris instead of glycine-Tris as in the usual (Laemmli)-SDS-PAGE. Gels were poured with an acrylamide/bisacrylamide concentration of 10 % (v/v), by mixing appropriate volumes of Rotiphorese 30, gel buffer and ddH$_2$O and adding TEMED and APS solution. The stacking gel was poured accordingly with 4.5 % (v/v) acrylamide/bisacrylamide. Running conditions were the same as described for the SDS-PAGE.

Tricine-LDS-PAGE

When samples containing *b*-type cytochromes were applied to in-gel heme staining (see section 2.9.6), Tricine-LDS PAGE was performed as it is generally known to be better suited for this approach. The PAGE differs from the one mentioned before by the fact that SDS is replaced by LDS. In contrast to the other electrophoreses, Tricine-LDS PAGE was performed at 4 °C.

Sample treatment

When a protein sample was applied to PAGE for subsequent protein visualization by Coomassie staining or for Western blotting, the ready-made loading buffer RotiLoad 1 (Roth, Karlsruhe, Germany) which contains mercaptoethanol as reductant, was used. When gels were run for subsequent heme staining, the following non-reducing loading buffer was used:

2 × non-reducing loading buffer:	125 mM Tris, 10 % (w/v) SDS, 20 % (w/v) glycerol, 0.005 % (w/v) bromophenol blue

For heme staining in Tricine-LDS-PAGE an appropriate sample buffer was used were LDS was used instead of SDS. Usually the samples were incubated at 100 °C for 5 min after the addition of loading buffer. However, when DsrM and DsrP were applied to gel electrophoresis, the samples were incubated at RT for 30 min as membrane proteins can irreversibly aggregate in SDS at temperatures >50 °C (Schägger, 2006).

2.9.6 Protein visualization and identification

Coomassie staining

Staining solution: 50 % (v/v) methanol
10 % (v/v) acetic acid
40 % (v/v) dH$_2$O
0.25 % (w/v) Coomassie Brilliant Blue R 250

Destain solution: 20 % (v/v) methanol
10 % (v/v) acetic acid
70 % (v/v) dH$_2$O

After protein electrophoresis the stacking gel was removed, and the running gel was transferred into Coomassie staining solution. After at least 30 min incubation the staining solution was replaced by destain solution. The latter was exchanged several times, until the protein bands were clearly visible. The gel was conserved by vacuum drying on an Aldo-Xer gel dryer (Schütt (Göttingen, Germany)).

Heme staining

heme staining solution: 4.5 mg TMBZ dissolved in 15 ml methanol
+ 35 ml 250 mM sodium acetate, pH 5.0

In-gel heme staining was carried out by the method of Thomas et al. (1976). After PAGE the stacking gel was removed and incubated in the heme staining solution for 30 min in the dark with gentle agitation. Then 600 µl of a 30 % H$_2$O$_2$ solution were added and the gel was further incubated until signals appeared. The gel was washed several times with dH$_2$O and scanned.

2.9.7 Western blotting

Towbin blotting buffer: 1.52 g Tris, 7.2 g glycine, 100 ml methanol,
ad 500 ml dH$_2$O

Proteins were transferred from PAGE gels onto Protran BA 85 cellulose nitrate membranes (Schleicher & Schuell (Dassel, Germany)) using a Transblot SD Semi-Dry transfer cell (Bio-Rad (München, Germany). After electrophoresis the stacking gel was discarded and the resolving gel was transferred to Towbin blotting buffer and incubated for 15 min. A gel-sized nitrocellulose membrane was also incubated in Towbin blotting buffer for at least 20 min. For the transfer, three layers of 3MM Chromatography paper (Whatman, Maidstone, United Kingdom) were soaked in blotting buffer and placed onto the anode. The membrane was placed onto the paper followed by the gel and another three layers of buffer soaked 3MM Chromatography paper. Finally the cathode was connected on top of the layers and the protein transfer was performed at 15 V for 15 - 40 min depending on the proteins molecular weight. Membranes were used either for immunological protein detection (see section 2.9.8) or Ponceau staining (see section 2.9.9).

2.9.8 Immunological protein detection

TBS: 50 mM Tris, 150 mM NaCl, pH 7.4

The immunological detection of proteins after the transfer onto a nitrocellulose membrane was performed using the primary antibodies listed in Table 2.2 and the appropriate secondary antibodies. The His-tag antibody was used according to the manufacturer's instructions (Merck, Darmstadt, Germany)
For all other antibodies, the procedure was the following:
Directly after Western transfer the membrane was incubated over night in TBS with 5 % (w/v) skim milk at 4°C. After 1 h incubation at room temperature the membrane was washed five times with TBS. The primary antibody was applied at a concentration of 1:1000 in 20 ml TBS with 0.1 g BSA for at least 3 hours. After several washing steps, an anti-rabbit antibody with a coupled horse radish peroxidase was used in a dilution of 1:5000 in 20 ml TBS with 0.1 g BSA and incubated for 1 h. The blot was washed twice and the detection was either accomplished using 4-chloro-1-naphthol or the SuperSignal West Pico chemiluminescent substrate. In the first case, the membrane was transferred into 43 ml dH$_2$O and 7 ml ethanol containing 30 mg 4-chloro 1-naphthol were added. The reaction was started by the addition of 20 µl H$_2$O$_2$. After bands were visible the blot was washed several times and documented. The detection using the SuperSignal West Pico chemiluminescent substrate was carried out according to the manufacturer's instructions (Thermo Fisher Scientific, Waltham, USA).

2.9.9 Ponceau staining

Ponceau S staining solution: 0.1 % (w/v) Ponceau S in 5 % (v/v) acetic acid

Ponceau staining is a reversible staining method that can be applied to proteins that were transferred onto cellulose nitrate membranes. The membranes were first incubated in the staining solution for 1 hour with gentle agitation. Then the staining solution was removed and the membranes were washed with dH_2O to remove the background.

2.9.10 Buffer exchange and protein concentration

Removal of undesired compounds such as imidazole or salt from a protein sample as well as buffer exchange was either carried out by dialysis or by gel filtration.

For dialysis the protein was filled into prepared dialysis tubes that were placed in a jar holding a large volume of appropriate buffer. The volume of buffer was at least fold the volume of the protein sample. Dialysis was carried out overnight at 4 °C with gentle stirring followed by another dialysis against 1 – 1.5 l buffer under the same conditions.

For desalting by gel filtration HiTrap Desalting columns were used according to the manufacturer's instructions (GE Healthcare, Uppsala, Sweden).

Dependent on the volumes of the protein samples, the samples were concentrated using either Amicon Ultra centrifugal filters (Millipore) with a molecular weight cutoff of 3 or 10 kDa or Vivaspin 500 centrifugal filters (Sartorius, Göttingen, Germany) with a 5 kDa MW cutoff membrane.

2.9.11 Analysis of protein interactions by coelution assays

To investigate protein-protein interaction between DsrK and DsrC, coelution assays were carried out taking advantage of the Strep-Tag fused to the C-terminus of DsrK. 10 pmol DsrC and 8.1 pmol DsrK in a final volume of 2 ml were incubated at room temperature for 30 minutes with gentle agitation. The mixture was then applied to a Strep-Tactin superflow column (IBA BioTagnology, Göttingen, Germany) with a column volume of 0.5 ml followed by purification according to the manufacturer's instructions. A control experiment was done under the same conditions with the exception that DsrK was omitted in sample that was applied to the Strep-Tactin column.

2.9.12 Quinone interaction

In order to test a putative interaction of the cytochromes with quinones, reduction assays with menadione were performed inside of the anaerobic chamber. Menadione is an analogue of menaquinone which lacks the isoprenoid side chain and which is therefore water soluble. 0.1 mM menadione was first reduced to menadiol by the use of a 2-fold molar excess of sodium borohydride. UV/Vis spectroscopy was used to verify complete reduction of menadione to menadiol. Menadiol was directly added to the protein for the desired reduction assays. After 2 min a spectrum was recorded to monitor the reduction state of the cytochrome. As a control experiment a 50-fold molar excess of sodium borohydride was added to the protein. Sodium dithionite was used to achieve full reduction of the hemes.

2.9.13 *In vitro* reconstitution of FeS clusters

A 50-fold molar excess of DTT was added to the purified protein and 7-fold molar excesses of $FeCl_2$ and Na_2S were added dropwise. The mixture was incubated over night with gentle stirring at 6 °C. Insoluble material was removed by centrifugation and the protein was applied to a HiTrap desalting column to remove excess iron and sulfide.

2.9.14 Mass spectrometry

Mass spectrometry was carried out by Ana M. V. Coelho in the Mass Spectrometry Laboratory, Analytical Services Unit, Instituto de Tecnologia Química e Biológica (ITQB), Universidade Nova de Lisboa. 5 µl of protein samples were desalted, concentrated and eluted using R1 (RP-C4 equivalent) microcolumns. Proteins were eluted directly onto the MALDI plate with sinapinic acid (10 mg ml^{-1}) using 50 % (v/v) acetonitrile and 5 % (v/v) formic acid. Mass spectra were acquired in the positive linear MS mode using a MALDI-TOF/TOF, Applied Biosystems, model 4800 (PO 01MS).

2.10 Protein biophysical methods

The biophysical methods applied in this work revealed significant results for the characterization of the proteins. In order to aid the understanding of the results and their interpretation, the theoretical background of these methods is briefly introduced.

2.10.1 UV/Vis absorption spectroscopy

Theoretical background

In UV/Vis absorption spectroscopy (in the following UV/Vis spectroscopy) the absorption of light by a sample is recorded in the ultra violet (UV) and in the visible (Vis) region. This region includes light with wavelengths approximately between 100 and 800 nm. In spectrophotometer a light beam is sent through the sample and the absorbance of the solution increases as attenuation of the beam increases. Absorbance (E) is directly proportional to the path length (d), and the concentration (c) of the absorbing species as is stated by Beer's law:

$$E = \varepsilon \times c \times d$$

Due to this, UV/Vis spectroscopy is often used for the quantification of samples with a known extinction coefficient (ε). Absorption of light occurs when electrons are promoted from their ground state to an excited state. The energy that is needed for that is provided by a light of a distinct wavelength which is accordingly missing in the spectrum recorded afterwards. The difference in the energy level between the ground state and the excited state depends amongst other upon the shape of the molecule orbital. In cytochromes, this shape is dependent on several factors (e.g. porphyrin molecule, redox state of the iron, iron ligation, spin state of the iron). There are transitions occurring in the distinct orbitals of the porphyrin molecule but also interactions between the orbitals of the porphyrin ring and the central metal ion occur, giving rise to absorption in the UV/Vis spectrum.

A classical UV/Vis spectrum of a cytochrome is characterized by at least four bands due to the heme and another band due to the absorption of the protein component. The four bands originating from the heme are referred to as δ, γ (or Soret), β and α. While the δ band is only visible in ferric cytochromes, β and α peaks are only observed in ferrous heme proteins. The characteristic of the Soret peak is that it shifts upon reduction of the heme iron.

Accomplishment

UV/Vis spectroscopy at room temperature was performed using an Analytic Jena Specord 210 spectrophotometer. Inside of the anaerobic chamber, spectra were recorded using a Perkin Elmer Lambda 11 spectrophotometer (Perkin Elmer, Waltham, USA). Quarz cuvettes were used in each case. Low temperature spectra were recorded with an OLIS DW-2 split-beam spectrophotometer (Olis, Bogart, USA). In this case the sample was filled into a support that was mounted within the light beam of the photometer in a reservoir filled with liquid nitrogen.

2.10.2 Redox titrations

Theoretical background

Redox titration by UV/Vis spectroscopy is a spectroelectrochemical method that can be used to investigate the redox properties of cytochromes. As already mentioned in section 2.10.1, the absorbance spectra of ferric and ferrous cytochromes differ. Upon reduction, the Soret peak shifts to a longer wavelength and α and β peaks become visible in a smooth transition. The Soret and the α peak are best suited for the observation of the redox state due to their intensity.

Accomplishment

In UV/Vis redox titrations the cytochrome is reduced stepwise by the addition of an appropriate reductant. The redox potential and an absorption spectrum are recorded in each step. After processing of the data, the relative absorption is plotted against the potential and the obtained plot is simulated using the Nernst equation.

All redox titrations were performed inside of an anaerobic chamber. 2.5 ml of a protein sample containing 1-3 µM heme was filled into a cuvette. A little stir bar was inserted and the cuvette was sealed with a rubber and placed inside of the Perkin Elmer Lambda 11. An InLab redox micro electrode (Mettler-Toledo, Schwerzenbach, Switzerland) was inserted into the cuvette through a hole in the rubber and the electrode was connected to a Lab850 pH/mV meter (Schott, Mainz, Germany). Finally the still opened spectrophotometer was protected from light and a first spectrum was recorded from 350 to 700 nm. Appropriate redox mediators covering the range of the investigated potential (Table 2.7) were added in a final concentration of 0.8 - 1.2 µM. Redox mediators are small electroactive compounds that effectively shuttle electrons between the analyte and electrode. The mixture was allowed to equilibrate with gentle stirring until the recorded potential was stable (30 – 60 min). The titration was carried out by adding small volumes (0.5 – 2 µl) of dithionite stock solutions (12.5 – 200 mM) using a Hamilton syringe (Hamilton, Bonaduz, Switzerland). When the dithionite was added, the potential was allowed to equilibrate and recorded. In parallel an absorption spectrum was recorded. This procedure was repeated until no further changes were observed in the UV/Vis spectrum. For data interpretation, the absorbance values were picked at distinct wavelengths from each spectrum. This was first the wavelength that was used to monitor the reduction state of the cytochrome (i.e. the wavelength of the Soret band or the α band), which is called wavelength X. Furthermore the wavelengths of the isosbestic points were picked for the correction of the values. Isosbestic points of a cytochrome are the specific wavelengths where the absorbance of the cytochrome in the oxidized an in the reduced state is identical. Two isosbestic points were chosen with a lower and a higher wavelength, respectively, compared to wavelength X. An imaginary line was drawn between the two isosbestic points and the distance between this line and the absorbance at wavelength X was calculated. The distance was normalized between 0 and 1 and the values plotted against the recorded potentials that were referenced to the standard hydrogen electrode.

Table 2.7: Redox mediators used in this study; E_0 – midpoint redox potential

Redox mediator	E_0 [mV]
Dichlorophenolindophenol	+217
1,2 naphtoquinone	+180
Trimethylhydroquinone	+115
Duroquinone	+50
Indigotrisulfonate	-70
2 hydroxy-1,4-naphtoquinone	-145
Anthroquinone-2-sulfonate	-225
Neutralred	-325
Bezylviologen	-360
Methylviologen	-446

Finally, the obtained plot was simulated by adding the adequate number of hemes with different potentials to the Nernst equation. This equation is used in electrochemistry to describe the potential of a redox pair dependent on the concentration.

$$E = E_0 + \frac{0.059\ V}{z_e} \lg \frac{c_{ox}}{c_{red}}$$

E: reduction potential

E_0: standard reduction potential

z_e: the number of transferred electrons

c_{ox}: concentration of the oxidized species

c_{red}: concentration of the reduced species

2 Methods

In practice, individual Nernst equations were calculated for each assumed heme with potentials in the range of -500 mV to + 500 mV and the simulation performed by calculating the average for each potential.

2.10.3 EPR spectroscopy

Theoretical background

Electron paramagnetic resonance (EPR) spectroscopy is a technique for the detection of paramagnetic compounds (i.e. compounds that have unpaired electrons). It is based on the effect that electrons align accordingly to their spin state either parallel or antiparallel relative to an external magnetic field (Zeeman effect). An unpaired electron can move between the two energy levels by emitting or absorbing energy. Absorption occurs, when the energy difference corresponds to the magnetic field. The EPR spectrum, which represents the first derivative of the absorption spectrum, is recorded by applying a constant energy and varying the magnetic field. The g values (g_x, g_y, g_z also referred to as g_{max}, g_{med} and g_{min}) are characteristic for the features of the sample (see below).

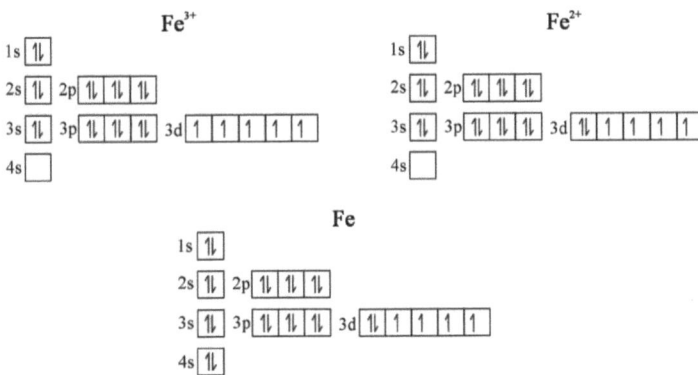

Figure 2.1: Electron configuration of iron atoms at different valency states.

The transition metal iron has the electron configuration of $1s^2 2s^2 2p^6 3s^2 3p^6 3d^6 4s^2$ in its uncharged state. Distribution of electrons within the electron configuration model is explained by the Aufbau principle which includes Hund's law saying that every orbital in a subshell is singly occupied with one electron before any orbital is doubly occupied, and that all electrons in singly occupied orbitals

2 Methods

have the same spin. In Figure 2.1 it can be seen that in each valency state shown (Fe, Fe^{2+}, Fe^{3+}), the iron has unpaired electrons and is therefore paramagnetic and accordingly EPR active.

EPR spectroscopy is a powerful technique for the analysis of cytochromes and FeS clusters. It is well suited to analyze the redox state, the ligation and the spin state of the iron atom as is explained in the following. The three-dimensional molecule orbital where the unpaired electrons are located, determines the shape of the EPR signal. The molecule orbital in turn depends on the structure of the complex, which has therefore a direct influence on the EPR signal (Figure 2.2).

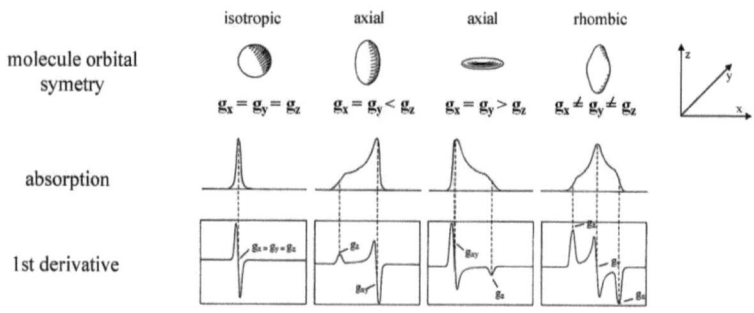

Figure 2.2: Origin of basic EPR spectra. Modified after Palmer (2000).

Six-coordinated hemes are octahedral complexes and give usually rise to a rhombic EPR signal when the axial ligands are not perpendicular. The axial ligands and the redox state of the iron influence the form of the molecule orbital leading to an EPR spectrum with characteristic g values. The spin state of the heme iron can be understood by regarding the fact that Hund's law is only true for degenerate orbitals (i.e. orbitals that have the same energy level). However, in octahedral complexes, splitting of the d orbitals occurs into different energy levels. Now two situations are feasible: i) the energy cost (ΔE) to place an electron into an already singly occupied orbital is lower than the energy cost to place the electron into the orbital with a higher energy level. In this case the total electron spin of the complex is lower and the complex is referred to as low-spin heme. ii) the energy cost (ΔE) is lower to place the electron in the orbital with the higher energy level. In this case, the complex is referred to as high spin complex. Figure 2.3 shows the electron configuration of the d orbitals in ferric and ferrous complexes of high and low spin hemes. It can be seen that low spin ferrous heme compounds are diamagnetic and therefore EPR silent.

2 Methods

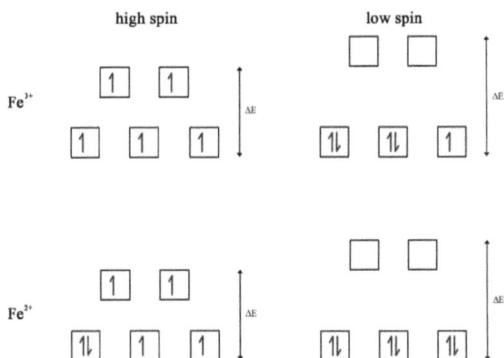

Figure 2.3: Electron configuration in degenerate d orbitals in ferric and ferrous octahedral complexes. See text for explanation.

Accomplishment

Electron paramagnetic resonance (EPR) spectroscopy was accomplished by Inês A. C. Pereira and Sofia S. Venceslau from the Laboratory for Bacterial Energy Metabolism of the Instituto de Tecnologia Química e Biológica (ITQB), Universidade Nova de Lisboa. Spectra were recorded using a Brucker EMX spectrometer equipped with an ESR 900 continuous-flow helium cryostat from Oxford Instruments.

2.10.4 Resonance Raman spectroscopy

Theoretical background

Raman spectroscopy is based on the effect that atom bonds are vibrating instead of being rigid. When a laser beam interacts with a vibrating bond, the photons either lose or gain energy resulting in a shift of laser beams wavelength. Scattering occurs accordingly to the different vibrating bonds in a molecule. Resonance Raman (RR) spectroscopy is an enhanced method in which the wavelength of the exciting light coincides with the wavelength of an electronic transition of the sample. This leads to enhanced scattering intensities and makes RR spectroscopy an ideal technique for studying heme proteins. The vibrations in the porphyrin ring strongly depend on the redox state, the ligation and the spin state of the heme iron. For instance a weak ligand or a five-coordinated high spin iron leads to a domed porphyrin molecule with significantly altered vibrations (Spiro, 1975). In the RR spectrum, marker bands can be found that are characteristic for the different states (ligation, redox, spin) of the iron atom.

Accomplishment

RR spectra were recorded by Smilja Todorovic either in the Laboratory for Raman Spectroscopy of Metalloproteins at the ITQB, or in the laboratory of Peter Hildebrandt at the Department for Chemistry of the Technical University of Berlin. Measurements were performed with a confocal microscope coupled to a Raman spectrometer (Jobin Yvon U1000). Samples were placed in a quartz rotating cell, excited with either 413 nm or 647 nm line from a krypton ion laser (Coherent Innova 302) and measured with 5 mW laser power and accumulation times of 60 s at room temperature. After polynomial background subtraction, the positions and line-widths of the Raman bands were determined by component analysis (Todorovic et al., 2006).

2.11 Bioinformatic methods

Sequence alignments were carried out using the clustal W algorithm (Larkin et al., 2007). Amino acid sequences were compared with the database using the Basic Local Alignment Search Tool (BLAST) (Altschul et al., 1990). Transmembrane helices were predicted using MEMSAT-SVM (Nugent & Jones, 2009). The helical wheel projection applet of the University of Virginia (Charlottesville, USA) was used for helical wheel projections. The applet is available at: http://cti.itc.virginia.edu/~cmg/Demo/wheel/wheelApp.html.

AMPHIASEEK was used for the prediction of amphiphatic α helices (Sapay et al., 2006).

The prediction of signal peptides and of cleavage positions was conducted using the SignalP and the LipoP software (Bendtsen et al., 2004; Juncker et al., 2003).

3 RESULTS

3.1 Enrichment and initial characterization of DsrMKJOP from *A vinosum*

3.1.1 Construction of mutants for the enrichment of DsrMKJOP

Two different strategies were applied to enrich DsrMKJOP from *A. vinosum* membranes:

Construction of *Allochromatium vinosum* Mhis2

In a first approach chromosomal *dsrM* was modified by the attachment of a His-tag coding sequence. Expression of the recombinant protein from the chromosome was thought to facilitate correct assembly of the membrane complex due to unaltered expression conditions. To maintain the distance to the original ribosome binding site, the His-tag coding sequence was inserted immediately after the start codon of the chromosomal *dsrM* by geneSOEing (see section 2.7.2). Therefore, primers dsrMHis-forwA and dsrMHis-revA were used for a first PCR and primers dsrMHis-forwB and dsrMHis-revB for a second. The two products served as templates for a third PCR that was carried out using primers dsrMHis-forwA and dsrMHis-revB. Since the mutation did only have a minimal effect on the size of the *dsrM* gene in comparison with the wild type, a *Kpn*I restriction site was introduced to facilitate the detection of the mutation by PCR and subsequent restriction assays.

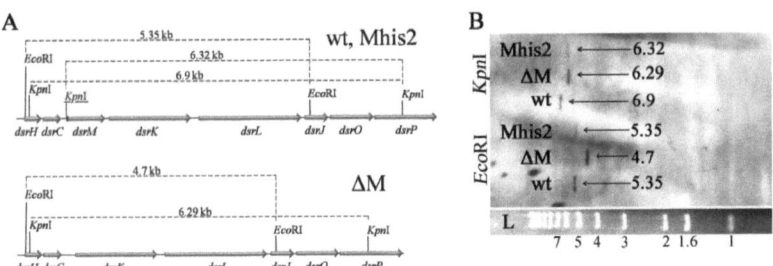

Figure 3.1: *A. vinosum* Mhis2 mutant confirmation; (A) schematic presentation of the genome of *A. vinosum* Rif50 (wt), Mhis2 and ΔM; the underlined *Kpn*I restriction site is only present in Mhis2; (B) Southern blot analysis; numbers correspond to the size of the molecule in kb; (L) 1 kb ladder;

The fragment was cloned into pK18mobsacB, sequenced and introduced into the genome of *A. vinousm* ΔM by the use of this plasmid. Thus, the wild type genotype was restored with the exception that a His-tag was fused to the N-terminus of the encoded DsrM. The mutant was termed *A. vinosum* Mhis2. The correct genotype of the mutant was verified by colony PCR and subsequent restriction with *Kpn*I as well as by Southern blot analysis (Figure 3.1). The DNA probe used for the Southern blot analysis was amplified from chromosomal *A. vinosum* Rif50 DNA using primers cvseq16F and PO-SK 2 that bind within the *dsrL* gene.

Construction of *A. vinosum* Δ*dsrJ*+Jhis

In a second approach a plasmid based expression of the periplasmic DsrJ including a C-terminal His-tag was carried out to facilitate purification of DsrMKJOP. For that purpose, *dsrJ* was amplified from the plasmid pETJHisSP including its native signal peptide and the C-terminal His-tag. Primers fwdsrJanker and JhisBam-rev were used introducing *Nde*I and *Bam*HI restriction sites, respectively. These restriction sites were used to insert the respective fragment downstream of the *dsr* promoter region in pBBR1MCS2-L (Lübbe et al., 2006) where it replaced the *dsrL* gene. The resulting plasmid was termed pBBRJHis and transferred from *E. coli* S17-1 to *A. vinosum* Δ*dsrJ* by conjugation as described in section 2.7.7.

3.1.2 Purification of DsrMKJOP

The enrichment of DsrMKJOP from *A. vinosum* Mhis2 and *A. vinosum* Δ*dsrJ*+Jhis was carried out under identical conditions: The *A. vinosum* mutant strains were grown in RCV medium in 1 l-flasks. Five days old cultures were induced with 2 mM sulfide and harvested after 3 hours. For purification, cells were resuspended in buffer and lysed by sonication. Insoluble components were removed by centrifugation and the membrane fraction was prepared from the supernatant by ultracentrifugation at $145\ 000 \times g$ for 3 h at 4 °C. Membranes were solubilized at least twice with 2 % dodecyl maltoside (DDM) (w/v) with gentle stirring on ice for several hours or overnight followed by another ultracentrifugation. The supernatant was supplemented with 15 mM imidazole and applied to a Ni-NTA agarose in batch mode. The column was washed with a stepwise gradient from 20 mM to 80 mM imidazole in lysis buffer supplied with 0.1 % DDM and the protein was eluted with 250 mM imidazole in lysis buffer. The eluted protein was concentrated to the desired concentration and desalted by gel filtration.

3 Results

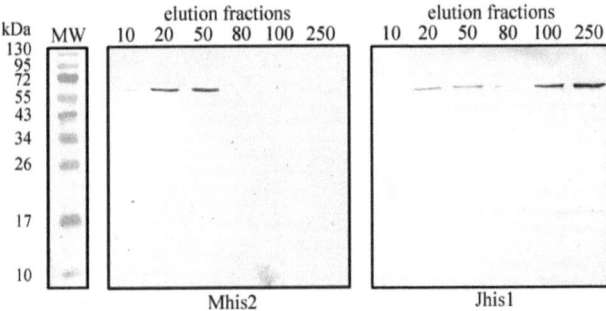

Figure 3.2: Western Blot for the detection of DsrK in the elution fractions of affinity chromatography using anti-DsrK antibody; solubilized membranes prepared from *A. vinosum* Mhis2 (left) and Jhis1 (right) were subjected to Ni NTA agarose columns and proteins eluted with a step gradient from 10 to 250 mM imidazole; the numbers on top of the blot indicate the concentration of imidazole in the respective fraction in mM; (MW), prestained molecular weight marker.

An anti-DsrK antibody was used to analyze the fractions that were eluted from the Ni-NTA column as this antibody gave an intense signal and a low background. As shown in Figure 3.2, DsrK eluted mainly in the wash fractions with 20 mM and 50 mM imidazole when the mutant Mhis2 was used while it eluted in the 100 mM and 250 mM elution fraction when mutant Δ*dsrJ*+Jhis was used. Accordingly the preparation that was received from this mutant was more pure. In the complementation assays (see section 3.3) it was shown that the expression of *dsrJ* from a plasmid under control of the *dsr* promoter had no negative effect on the phenotype of the mutant ruling out that this kind of expression was detrimental to the composition of the DsrMKJOP complex. For the following experiments DsrMKJOP was therefore enriched from *A. vinosum* Δ*dsrJ*+Jhis.

3.1.3 Characterization of DsrMKJOP

Western blot analysis

Detection of the proteins assembling the transmembrane complex was achieved by Western blot analysis. DsrM, DsrK, DsrJ and DsrO were readily detected in the preparation while a specific antibody against DsrP was not available (Figure 3.3). The theoretical mass of DsrM is 27.9 kDa, however, it appeared to be smaller upon SDS-PAGE analysis. Anomalous migration of DsrM has also been observed for the corresponding protein from *Archaeoglobus fulgidus* (Mander et al., 2002) and is generally observed in membrane proteins (Rath et al., 2009). DsrK was detected above the 55 kDa band of the molecular weight marker, while DsrO ran slightly below the 26 kDa band (Figure 3.3). This is in accordance with the calculated value of the processed protein while the non-processed protein would have a molecular mass of 28.9 kDa. It can therefore be concluded that in *A. vinosum*, the signal peptide of DsrO is cleaved off as previously reported (Dahl et al., 2005).

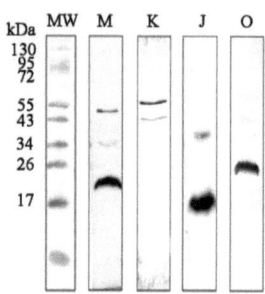

Figure 3.3: Western blot analysis of DsrMKJOP enriched from *A. vinosum* Δ*dsrJ*+Jhis using anti-DsrM (M), anti-DsrK (K), anti-DsrJ (J) and anti-DsrO (O) antisera after protein separation in 10 % Tricine-SDS-PAGE; (MW), prestained molecular weight marker.

A. vinosum DsrJ was identified via the anti-DsrJ antiserum and the according signal was detected slightly above the 17 kDa band of the molecular weight marker (Figure 3.3). To ascertain if the signal peptide of DsrJ is cleaved off, the molecular mass of DsrJ was judged by SDS-PAGE. An unstained molecular weight marker served as reference in this experiment (Figure 3.4). DsrJ was found to have a molecular mass of ~19 kDa which is in perfect accordance with the mass predicted for the His-tagged *c*-type cytochrome including its signal peptide while the calculated molecular mass of the processed form would only have been 16.6 kDa. Additionally, a soluble version produced in *E. coli* was subjected to Tricine-SDS-PAGE as comparison. The PelB leader that was used instead of the native signal peptide is removed in this protein (see section 3.2). The soluble version lacking a signal peptide was found to be significantly smaller than the protein produced in *A. vinosum* (Figure 3.4). These results confidently revealed that the signal peptide of DsrJ is not cleaved off in *A. vinosum*. This is also corroborated by bioinformatic analyses of DsrJ (see section 3.2.1).

3 Results

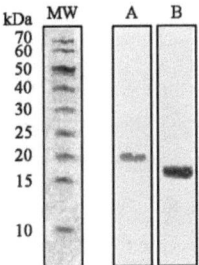

Figure 3.4: Western blot analysis after 15 % SDS-PAGE using anti-DsrJ antibody; (A), fraction enriched in DsrMKJOP from *A. vinosum*; (B), purified soluble DsrJ produced in *E. coli*; MW, molecular weight marker stained with Ponceau S.

Coomassie staining

In Coomassie stained Tricine-SDS-PAGE gels definite identification of each of the DsrMKJOP proteins was difficult. This was because the dominant bands were obviously contaminating proteins that were also present when *A. vinosum* wild type extracts were purified via affinity chromatography under identical conditions (Figure 3.5).

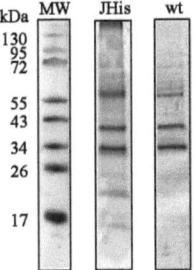

Figure 3.5: SDS-PAGE analysis of fractions after Ni-NTA affinity chromatography; fractions obtained from *A. vinosum* Δ*dsrJ*+Jhis (JHis) and *A. vinosum* wild type (wt) applied to 10 % Tricine-SDS-PAGE gels and subsequent Coomassie staining; (MW), prestained molecular weight marker.

One of the bands moreover reacted with an anti-His antibody (not shown) indicating that it is a naturally histidine rich protein from *A. vinosum*. It was not attempted to further purify DsrMKJOP

3 Results

from *A. vinosum* since this work was focussed on the individual production and characterization of the individual proteins in *E. coli*.

Heme staining

Heme staining of Tricine-SDS-PAGE gels identified DsrJ as the sole *c*-type cytochrome in the preparation that was enriched in DsrMKJOP (Figure 3.6A). When Tricine-LDS-PAGE gels run at 4 °C were subjected to heme staining the strong signals of DsrJ multimers and the resulting background prevented the detection of *b*-type cytochromes that usually give weaker signals due to the loss of heme groups during electrophoresis (Figure 3.6B). Furthermore the smear on top of the resolving gel gave a positive signal indicating that some amount of cytochomes did not enter the gel. In a control experiment *A. vinosum* wild type membrane extracts were subjected to nickel-chelate affinity chromatography. The eluate was separated by Tricine-SDS-PAGE and Tricine-LDS-PAGE run at 4 °C and the gels were subjected to heme staining. No signals were detected, thus excluding that cytochromes were present in this preparation (not shown).

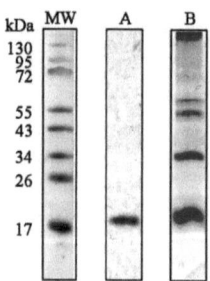

Figure 3.6: Detection of cytochromes in a fraction enriched in DsrMKJOP from *A. vinosum* Δ*dsrJ*+Jhis by in-gel heme staining after separation in 10 % Tricine-SDS-PAGE at rt (A) and after separation in 10 % Tricine-LDS-PAGE at 4 °C (B); (MW), prestained molecular weight marker.

UV/Vis spectroscopy

First characterization of the membrane fractions enriched in DsrMKJOP was carried out by UV/Vis spectroscopy (Figure 3.7). The spectrum of the sample is dominated by the intense Soret peak at 410 nm that shifts to 419 nm upon reduction. The spectrum showed no characteristic signals for the siroheme sulfite reductase revealing that the DsrMKJOP complex was successfully enriched from *A. vinosum* without co-purification of DsrAB. In the reduced state α and β peaks were observed at 554 and 522 nm, respectively. The α peak was very broad and the beta peak displayed a shoulder at 530 nm indicating that hemes *b* and *c* were simultaneously present in the preparation. The

3 Results

absorbance in the 430-500 nm region may partly be attributed to FeS clusters. The weak peaks at 485 and 590 nm may be due to a contamination by carotenoids that are ubiquitous in the membranes of purple sulfur bacteria. These peaks were also observed in the UV/Vis spectrum of *A. vinosum* wild type membrane extracts that were subjected to nickel-chelate affinity chromatography in a control experiment. As expected from the heme staining experiments, no heme compounds were detected in the UV/Vis spectrum of this preparation (not shown). This finally showed that - although the DsrMKJOP preparation was not completely homogeneous - the cytochromes of the complex fully accounted for the heme-specific features of the spectrum.

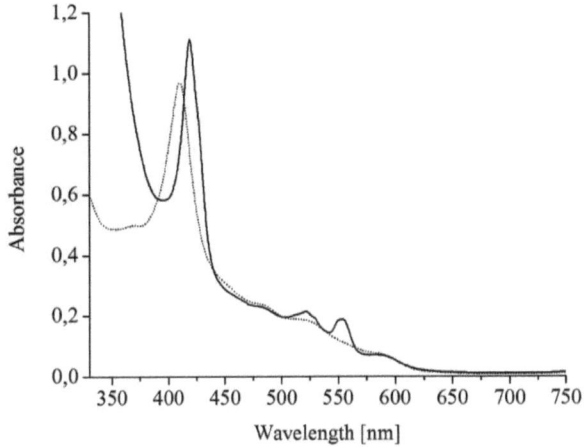

Figure 3.7: UV/Vis spectroscopy of the fraction enriched in DsrMKJOP from *A. vinosum* ΔdsrJ+Jhis in the oxidized state (dotted line) and in the reduced state (solid line).

Hemes *b* and hemes *c* differ in the wavelengths where their major peaks are observed in the oxidized as well as in the reduced state. The Soret peak of ferric hemes *c* are usually found between 408 and 410 nm while in the ferrous state, the α peak is located between 549 and 551 nm. The peaks of hemes *b* are shifted to longer wavelength. The Soret peak is usually found between 412 and 415 nm in the oxidized state. In the spectra of ferrous hemes *b*, the α peak is usually observed between 556 and 558 nm (Lemberg & Barrett, 1973). Especially the position and the intensity of the α and the Soret peaks can be used to assess the redox state of hemes in a mixture of different types of hemes although overlapping of these peaks limits this approach. The fraction enriched in DsrMKJOP was investigated for the redox state of the hemes at different potentials by monitoring the absorbance changes in the Soret and in the α peak by UV/Vis spectroscopy (Figure 3.8). When the experiment started, the Soret peak was localized at 410 nm and no α peak was observed, revealing that all hemes were oxidized at +220 mV. At +180 mV an α peak was observed at 560 nm indicating beginning reduction of the hemes *b*. Furthermore splitting of the Soret peak was observed at this potential: A small peak could be observed at 522.5 nm which corresponds to the Soret peak of the reduced hemes *b* while the maximum of the Soret peak resembling hemes in the oxidized state was slightly red-shifted to 408 nm. This can be explained by the fact that the ferric hemes *c* contribute to a higher extent to this peak at +180 mV as compared to the completely oxidized spectrum recorded at + 220 mV. Upon decreasing the potential to -50 mV, the hemes *b* were further reduced as can be seen by the increasing α band at 560 nm. The intensity at 560 nm did not increase upon further lowering the potential revealing that the hemes *b* were fully reduced at -50 mV. The beginning reduction of the hemes *c* was observed as an α peak appeared at 551 nm and the Soret peak was represented by a broad absorbance between 400 and 430 nm. The spectrum recorded at -240 mV revealed that the α peaks were fused to a broad peak with a maximum at 554 nm and that the Soret peak was fully shifted to 419 nm. When the potential was even lowered, no absorbance changes were observed showing that all hemes in the preparation were fully reduced at -240 mV. It was not attempted to determine the midpoint redox potentials of the individual hemes since different types of hemes (i.e. hemes *c* and hemes *b*) were present in the preparation, that contribute differently to the absorbance at the wavelength that is used for the calculation of the redox titration. Furthermore the hemes *c* in DsrJ differ in their axial ligation and therefore contribute differently to the absorbance spectrum. Besides this, there is most likely redox cooperativity between the hemes and between the different hemes and the FeS clusters that were present in the preparation as well.

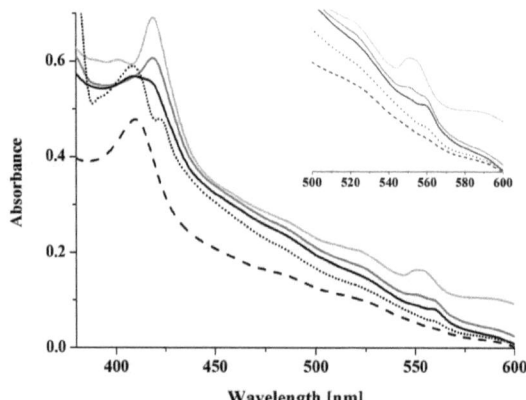

Figure 3.8: UV/Vis spectra of a preparation enriched in DsrMKJOP from *A. vinosum* Δ*dsrJ*+Jhis at various potentials: +220 mV (light gray); +180 mV (dark gray); -50 mV (black); -180 mV (dotted); -240 mV (dashed).

In another experiment the interaction of the DsrMKJOP complex with the soluble menaquinol analogue menadiol was analyzed. Under anoxic conditions menadiol was added to the protein sample and the redox state of the hemes was investigated by UV/Vis spectroscopy (Figure 3.9). After addition of menadiol the Soret peak revealed a little shoulder and a peak was observed at 560 nm, representing the α peak of the cytochromes *b* in the preparation. No change was observed in the region between 540 and 555 nm ruling out that the *c*-type cytochromes were reduced. The reduction was found to be specific since the negative control (addition of 50-fold molar excess of borohydride) did not lead to absorbance changes. For full reduction, dithionite powder was added. The results clearly show that some amount of heme *b* was specifically reduced by menadiol while the *c*-type cytochromes were not.

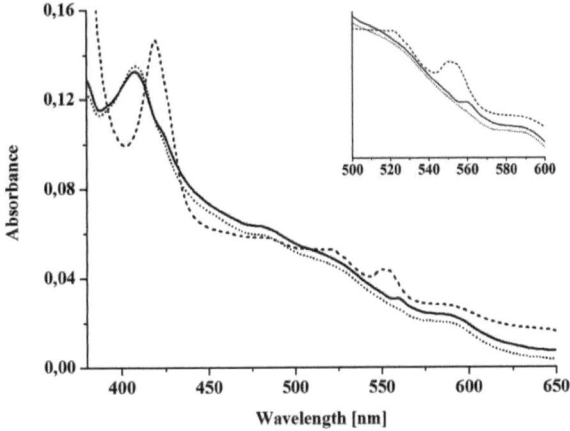

Figure 3.9: UV/Vis spectroscopy of the fraction enriched in DsrMKJOP from *A. vinosum* Δ*dsrJ*+J*his* in the oxidized state (dotted line), after addition of menadiol (solid line) and after addition of sodium dithionite (dashed line); the inset is a close up of the 500 – 600 nm region.

Heme quantification

Pyridine hemochrome spectra can be used to calculate the amount and type of hemes in a preparation containing different types of hemes (Berry & Trumpower, 1987). Pyridine serves as axial ligand to the hemes and the cytochromes therefore absorb at distinct wavelengths with a distinct absorption coefficient independently from their native axial heme ligands. From the pyridine hemochrome spectrum, hemes *b* and hemes *c* were calculated to be present in a 1:1.26 ratio in the preparation that was enriched in DsrMKJOP.

3.2 Individual production and characterization of DsrJ

3.2.1 Bioinformatic analysis of DsrJ

An N-terminal signal peptide for the translocation via the Sec pathway is predicted for DsrJ as mentioned before. Such a signal peptide is usually characterized by a short sequence of mainly positively charged amino acids followed by a hydrophobic region. Due to its hydrophobic character, this region is very often annotated as a transmembrane helix by convenient software programs. This is also true for DsrJ as various programs predict a transmembrane helix located at the N-terminus of DsrJ. A second mainly positively charged region is usually found in signal peptides behind the hydrophobic region which is followed by a cleavage site for the leader peptidase. Depending on the method that is used to predict the cleavage position of the signal peptide, the SignalP program (Bendtsen et al., 2004) predicts a most likely cleavage position between amino acids 21 and 22 or between 27 and 28. The latter cleavage position is predicted to be more likely by the LipoP software (Juncker et al., 2003). However, depending on the program and the method used, the overall cleavage site probability is only slightly above the cutoff. In this context it has to be noted that in some cases, the signal peptide of a protein is not cleaved off but serves as membrane anchor. This has for instance been reported for the cytochrome *c* subunit FccA from *Chlorobaculum thiosulfatiphilum* (formerly *Chlorobium limicola*) (Verte et al., 2002) and - more importantly - for the DsrJ homologue HmeE from *Archaeoglobus fulgidus* (Mander et al., 2002). As obvious from the Western blot results described in section 3.1.3, this is also true for DsrJ from *A. vinosum*.

3.2.2 Heterologous production of recombinant DsrJ

Various constructs were used for the expression of recombinant DsrJ in *E. coli*. In each case *E. coli* BL21 (DE 3) was first transformed with the pEC86 plasmid that provides the expression of the genes encoded in the *ccm* operon under aerobic conditions (Arslan et al., 1998) and then transformed with the appropriate plasmid used for expression of *dsrJ*. First trials in expression of *dsrJ* in *E. coli* have already been made but satisfying production of holoprotein was not achieved (Dahl et al., 2005; Schneider, 2007). Therefore, expression conditions were optimized for the production of DsrJ.

Proper maturation to a fully heme-loaded DsrJ holoprotein was absolutely dependent on slow production of the recombinant protein in *E. coli*. The required minimum expression level was achieved by the use of the lacT7 promoter in the pET22b vector and by completely omitting induction. 400 ml of NZCYM medium in a non-baffled Erlenmeyer flask were inoculated with a single colony of the appropriate clone and grown at 37 °C and 180 rpm agitation for 16 to 18 hours. These conditions led to deeply red colored cells.

Production of soluble and membrane bound DsrJ

In vector pETJHis a His-tag is fused to the C-terminus of DsrJ and the native signal peptide is replaced by the PelB leader that is provided by the pet22b vector (Schneider, 2007). The PelB leader is the signal peptide of the pectate lyase from *Erwinia carotovora* and is commonly used for production of periplasmic proteins in *E. coli*. When this vector was used for expression of DsrJ, the majority of the protein was found in the soluble fraction (Figure 3.10). Using the primers fwdsrJanker and LdsrJXho-r, *dsrJ* was amplified including its native signal peptide and the resulting fragment was cloned into pET22b resulting in plasmid pETJHisSP. When this plasmid was used for production of DsrJ in *E. coli*, the protein was exclusively found in the membrane fraction (Figure 3.10). Furthermore the molecular weight was found to be higher in comparison to the version when the PelB-leader was used as signal peptide. These results clearly showed that the native signal peptide of DsrJ is not cleaved off but serves as membrane anchor in when the protein is produced in *E. coli*.

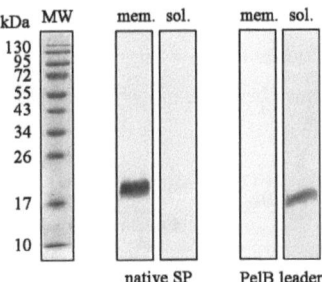

Figure 3.10: Localization of recombinant DsrJ in *E. coli*; DsrJ including the native signal peptide (native SP) and DsrJ carrying the PelB leader instead of the native signal peptide (PelB leader) applied to Western blot analysis using anti-DsrJ antibody; (mem.), membrane fraction; (sol.), soluble fraction; (MW), prestained molecular weight marker.

Production of Strep-tagged DsrJ

For the following experiments the soluble version of DsrJ was produced to facilitate purification and subsequent characterization. A Strep-tag fused to the C- terminus of DsrJ was used instead of the hexahistidine-tag to exclude possible interactions of additional histidines with the natural heme ligands. First, plasmid pPR-IBApelB was constructed by ligating the *pelB* leader sequence, which had been retrieved from pET22b using *Nco*I and *Xba*I, into pPR-IBA1. Primers LdsrJNco-f and

DsrJEco47III-r were used to amplify *dsrJ* from chromosomal DNA and the resulting fragment was inserted into pPR-IBApelB using the restriction enzymes *Nco*I and *Eco*47III. The resulting plasmid was termed pPRIBApelBJStrep. Production of proper holoprotein was not achieved with this construct. Therefore, the vector was digested with *Nco*I and *Hin*dIII and the insert was transferred back into the pET22b backbone resulting in the plasmid pETJStrep. This plasmid coded for soluble DsrJ including a C-terminal Strep-tag and provided a low basal expression due to the lacT7 promoter in pET22b in contrast to the T7 promoter in pPRIBApelBStrep. With vector pETJStrep production of proper holoprotein was achieved and 820 µg pure DsrJ were obtained from one liter expression culture.

Production of Strep-tagged DsrJC46S

To prove and to investigate the role of the cysteine heme coordination in DsrJ, a mutant form of the protein, DsrJC46S, was also produced. In this protein Cys46 was replaced with a serine. This amino acid substitution, which only replaces a sulfur atom with oxygen, was chosen to address the specific role of this sulfur atom without changing other stereochemical properties. The vector pETJC46SStrep was constructed in analogy to the vector pETJStrep with the exception that the insert was modified using GeneSOEing (**s**plicing by **o**verlap **e**xtension). Vector pETJHis was used as template for GeneSOEing. Primers T7-f and DsrJC46-r were used for the first and primers DsrJC46S-f and LpelBJHisXba-r were used for a second PCR. The obtained products were used as templates for another PCR that was carried out using primers LdsrJNco-f and LpelBJHisXba-r.

3.2.3 Purification of DsrJ and DsrJC46S

DsrJ and DsrJC46S were purified under identical conditions. For purification, the cells were thawed, resuspended in buffer W and lysed by sonication. The extract was centrifuged for 30 min at 25 000 × g at 4 °C to remove insoluble components. The supernatant was applied to an affinity chromatography column (Strep-Tactin Superflow). The column was washed and the protein eluted according to the manufacturer's instructions (IBA BioTagnology, Göttingen, Germany). The protein that was eluted from the Strep-Tactin column was deeply red colored and was used for EPR spectroscopy and mass spectrometry. SDS-PAGE analysis revealed two protein bands with molecular masses of ~ 16.5 and ~ 30 kDa, respectively. These masses correspond to the monomeric and dimeric form of processed recombinant DsrJ. Besides this, a third band resembling a protein with a molecular weight of ~ 12 kDa could be observed upon SDS-PAGE analysis. All bands gave positive signals in heme staining (Figure 3.11A) and reacted with anti-DsrJ and anti-Strep antibodies. Since the Strep-tag is fused to the C-terminus of DsrJ, it can be concluded that the ~12 kDa band corresponds to DsrJ that is shortened at the N-terminus.

3 Results

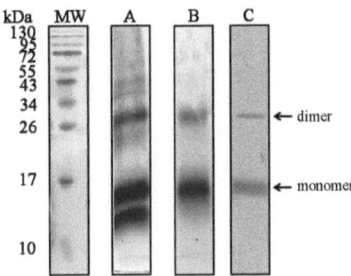

Figure 3.11: Purification of recombinant DsrJ; DsrJ eluted from the Strep-Tactin column (A) and eluted from the gel filtration column (B) applied to SDS-PAGE and subsequent in-gel heme staining; purified DsrJ in Coomassie stained SDS-PAGE gel (C); (MW) prestained molecular weight marker.

This may be due to wrong processing of the protein or proteolytic degradation. The use of protease inhibitors did not change the picture, excluding that proteolytic degradation occurred during the purification process. Dimerization of DsrJ was not impaired in the mutated protein, ruling out that a disulfide bridge is formed between Cys46 of two DsrJ molecules.

For use in heme quantifications, redox titrations or UV/Vis spectroscopy the protein was further purified by immediately applying the protein to a Superdex 75 pg gel filtration column that was equilibrated with 50 mM NaH_2PO_4, pH 7.0. Monomeric DsrJ eluted between 75 and 78 ml as seen in Figure 3.12 while the smaller artifact eluted between 81 and 85 ml. Using gel filtration chromatography DsrJ was purified to homogeneity as judged by SDS-PAGE analysis (Figure 3.11C).

3 Results

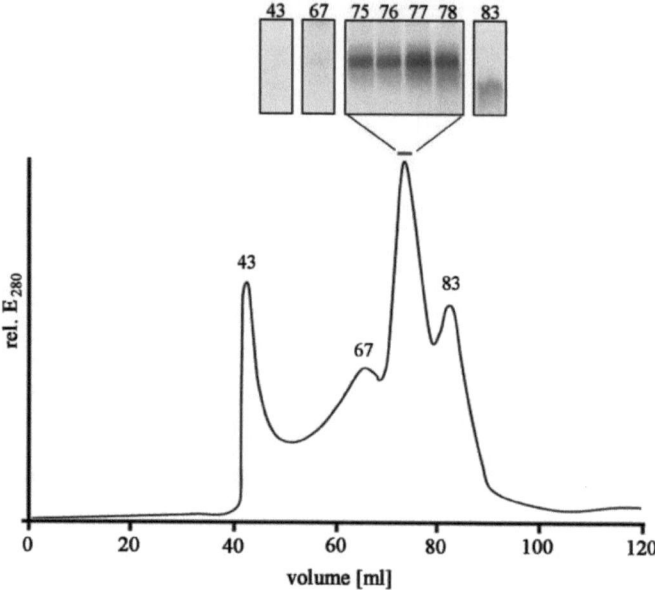

Figure 3.12: Chromatogram of the gel filtration using the Superdex 75 pg column; above the chromatogram a detail of a SDS-PAGE after heme staining is depicted; the numbers correspond to the elution volume of the respective fraction.

3.2.4 Characterization of recombinant DsrJ and DsrJC46S

Heme quantification

Heme quantification of the purified protein was carried out by pyridine hemochrome spectroscopy (Berry & Trumpower, 1987) and revealed 2.6 mol heme c per mol DsrJ for the wild type and 2.7 mol heme c per mol DsrJ for the mutated protein.

Table 3.1: Mass spectroscopy results of purified DsrJ and DsrJC46S.

Peak number	Observed mass [Da]		Theoretical mass [Da]	Mass difference [Da]
1	14888	DsrJ	14960.5	72.5
2	15495	DsrJ + 1 heme	15573	78
3	16109.2	DsrJ + 2 hemes	16185.5	76.3
4	16726	DsrJ + 3 hemes	16798	72
4b	16802.3	DsrJC46S + 3 hemes	16794	8.3

3 Results

Final proof that the holocytochromes were properly and completely loaded with heme was achieved by mass spectrometry. The accuracy of this experiment was quite low as the error (peak width at half height) for the individual peaks was found to be up to 82 Da. However, the difference between the theoretical and the obtained molecular masses was within this error range. Moreover, the difference between the individual observed masses was in the range of 607 to 616.8 Da, close to the theoretical mass of one heme (612.5 Da) (Table 3.1). The intensities of the peaks revealed that the vast majority of protein bound three hemes (16726 Da) while only minor peaks were detected for DsrJ with two (16109.2 Da), one (15495 Da) and zero (14888 Da) hemes bound, respectively. The preparation of DsrJC46S that was subjected to mass spectrometry revealed a major peak at 16802.3 Da. The observed mass corresponds to the mass of DsrJC46S including three hemes (Table 3.1). The low accuracy of the experiments did not allow the identification of a possible post-translational modification of the proteins.

UV/Vis spectroscopy

Purified DsrJ was characterized by UV/Vis spectroscopy. The protein as isolated was completely oxidized as addition of the strong oxidant hexacyanoferrate did not alter the spectrum (not shown). The spectrum of oxidized DsrJ is typical for low-spin ferric heme and revealed a Soret peak at 408 nm and a δ peak at 351 nm. Upon reduction the Soret peak shifted to 417.5 nm and α and β peaks were observed at 551 and 523 nm, respectively. An additional weak signal at 650 nm was detected in the oxidized and in the reduced state (Figure 3.13A). The UV/Vis spectrum of the purified mutated protein was unaltered as compared to that of original DsrJ (Figure 3.13B). The millimolar extinction coefficient ε_{408} of purified DsrJ in the oxidized state was calculated to be 251.4 $mM^{-1} cm^{-1}$.

3 Results

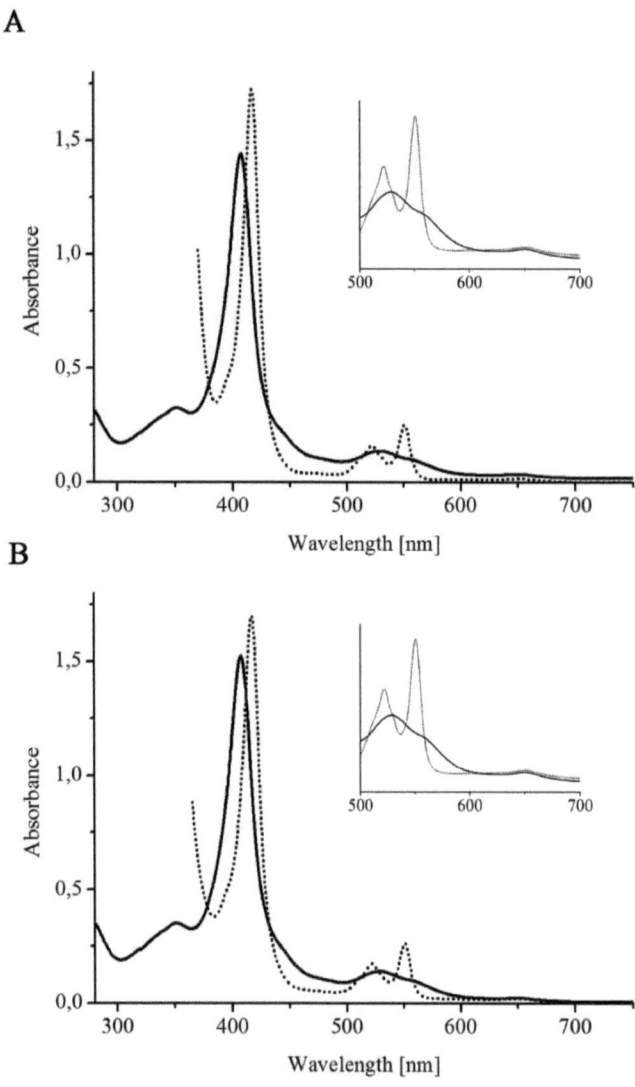

Figure 3.13: UV/Vis spectra of wild type DsrJ (A) and DsrJC46S (B) in the oxidized state (full line) and in the reduced state (dotted line); the inset is a closeup of the 500 – 700 nm region.

3 Results

EPR spectroscopy

The EPR spectra of purified DsrJ (Figure 3.14) are very informative concerning the spin states of the hemes as well as the axial ligation of the heme iron. High-spin heme complexes usually give rise to an intense signal with g = 6.0 in EPR spectroscopy. As shown in Figure 3.14A, no high-spin heme was observed in the EPR spectrum of wild type DsrJ. Besides the signal of non-specific iron at g = 4.3, the EPR spectrum revealed signals for low spin heme compounds at g = 2.961, 2.517 (g_{max}), 2.263 (g_{med}) and 1.858, 1.502 (g_{min}). The 2.961, 2.263 and 1.1502 values are in the range typically found for low spin hemes with His/His or His/Met ligation (Cheesman et al., 2001). The signals at 2.517 and at 1.858 are outside of this range since the lowest g_{max} value found for His/His, His/Met or His/Lys is 2.86 for a His/Met ligated heme in bacterioferritin from *Pseudomonas aeruginosa* (Cheesman et al., 1992). Values similar to the g_{max} = 2.517 and g_{min} = 1.858 have been observed for the His/Cys ligated hemes in SoxXA and PufC from *R. sulfidophilum* (Alric et al., 2004; Cheesman et al., 2001). This allowed a provisionally attribution of these values to a His/Cys ligated heme in DsrJ. A comparable signal was also detected in the DsrMKJOP complex from *D. desulfuricans* and proposed to be due to a potential His/Cys ligated heme in DsrJ (Pires et al., 2006).

However, spectral simulation of the two g_{max} = 2.961 and g_{max} = 2.517 species of wild type DsrJ indicates that the latter one accounts for substantially less than one heme (a maximum of ~ 0.5-0.6 heme) (S. S. Venceslau, I. A. C. Pereira, personal communication), suggesting that coordination of this heme is heterogeneous. The characteristic signal at 2.517 that was attributed to the His/Cys ligated heme disappeared in the EPR spectrum of DsrJC46S and the g_{min} = 1.858 significantly decreased (Figure 3.14B).

3 Results

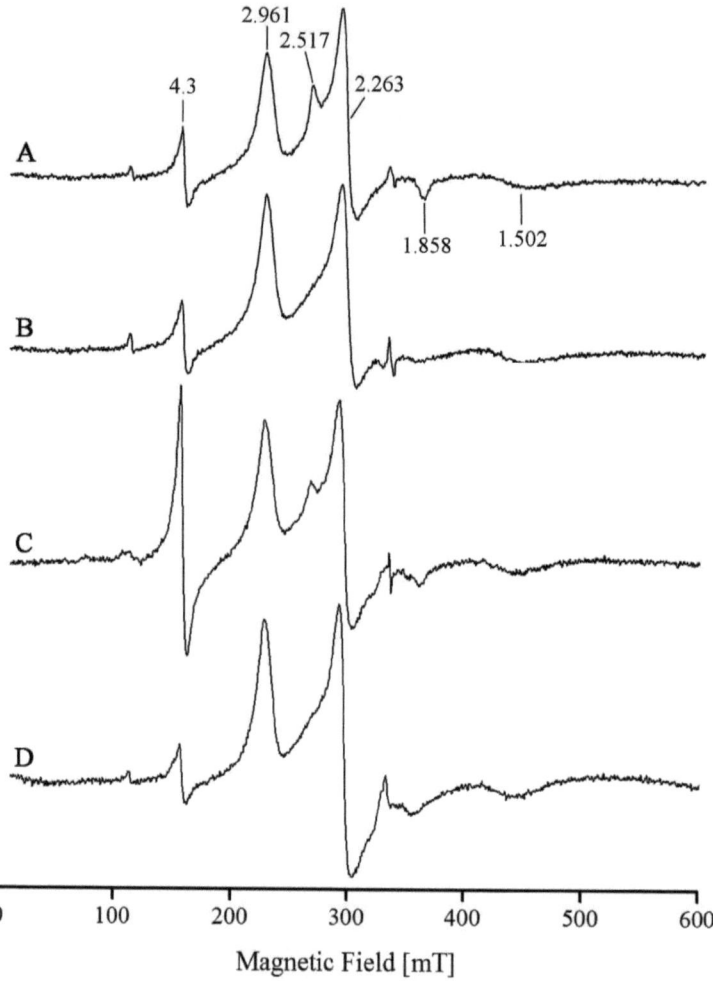

Figure 3.14: X-Band EPR spectroscopy of recombinant DsrJ; (A) 130 μM wild type DsrJ as prepared, the g-values of the most intense signals are labeled; (B) 180 μM DsrJC46S as prepared; (C) wild type DsrJ after reduction and re-oxidation; (D) DsrJC46S after reduction and re-oxidation. Experimental conditions: microwave frequency, 9.38 GHz; microwave power, 2 mW; modulation frequency, 100 kHz; and modulation amplitude, 1 mT; temperature, 10 K (A,B), 14 K (C,D).

However, the spectrum of reduced and re-oxidized DsrJC46S revealed a signal at ~1.8 (Figure 3.14D) indicating that this signal is most likely not due to a His/Cys ligated heme. Nevertheless the disappearance of the g_{max} = 2.517 strongly suggests that cysteine 46 is the ligand to one of the hemes in DsrJ. Upon reduction and re-oxidation of the wt protein, the intensity of the signal with g_{max} = 2.517 decreased, accounting to a maximum of 0.2-0.3 heme in the redox cycled protein. Comparing spectral simulations of the redox-cycled wt and mutant proteins the decrease in the g_{max} = 2.517 signal is in the order of 0.1 heme (S. S. Venceslau, I. A. C. Pereira, personal communication). This further indicates that the possible coordination of one of the hemes by Cys46 is only partial, and that ligand switching may occur during reduction.

When cysteine 46 is lost as axial ligand, one would expect the respective heme to become a 5-coordinated high-spin heme that can easily be detected in the EPR spectrum. However, a high spin heme is not observed in the DsrJ EPR spectrum, indicating that the heme iron is still 6-coordinated and that there is possibly another residue that can also coordinate this heme. This is corroborated by the fact that upon reduction and re-oxidation a decreasing g_{max} = 2.517 signal is compensated by an increasing g_{max} = 2.961. This verifies that the heme that was ligated by cysteine 46 is now in a ligation state that contributes to a g_{max} = 2.961 signal which may be a His/His or a His/Met ligation. The same holds true for the situation in DsrJC46S were a high spin heme is also not observed. Spectral integrations of the DsrJ and DsrJC46S EPR signals were carried out to quantify the increase of the g_{max} = 2.961 signal in comparison to the disappeared g_{max} = 2.517 signal. They showed indeed very similar relative intensity, which indicates that the loss of the g_{max} = 2.517 signal is completely compensated by an increase in the g_{max} = 2.961 signal (S. S. Venceslau, I. A. C. Pereira, personal communication).

In EPR spectroscopy, paramagnetic species (i.e. compounds that have unpaired electrons) can be detected. Ferric low spin hemes are paramagnetic while in the ferrous state they are diamagnetic and accordingly EPR silent (see section 2.10.3). A decreasing intensity of the EPR signal can therefore be observed upon reduction of a heme. Upon treatment with the mild reductant ascorbate, both the g = 2.961 signal and the g = 2.517 signal, that was attributed to the His/Cys ligated heme, decreased (Figure 3.15B). However, the relative contribution of the g_{max} = 2.517 signal seems to be larger than in the oxidized sample, suggesting that the heme responsible for this signal may have a lower potential than the others. Nevertheless, the partial reduction of this heme by ascorbate indicates that it is redox active and does not have an extremely low redox potential as observed for the His/Cys coordinated heme in SoxXA from *R. sulfidophilum* and *Paracoccus pantotrophus* (Cheesman et al., 2001; Reijerse et al., 2007) or DsrJ from *D. desulfuricans* (Pires et al., 2006). This conclusion was corroborated upon reduction with the strong reductant dithionite, which leads to complete reduction of the hemes as observed by EPR spectroscopy (Figure 3.15C).

3 Results

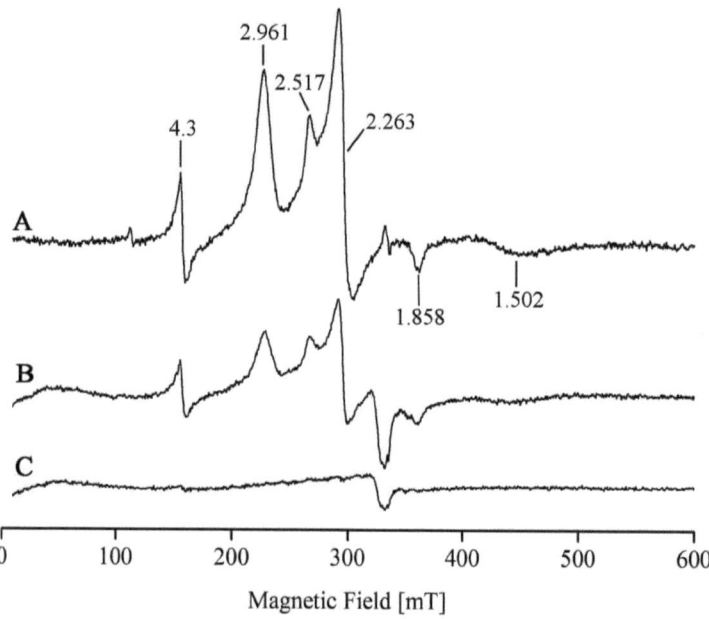

Figure 3.15: X-Band EPR spectroscopy of 130 µM wild type DsrJ at various reduction states; (A), as prepared; (B), after addition of ascorbate; (C), after addition of dithionite; Experimental conditions as described in Figure 3.14. The most intense signals are signals are labeled in (A).

Resonance Raman spectroscopy

The high frequency region (1300 – 1700 cm^{-1}) RR spectra of heme proteins, obtained upon Soret band excitation, are of special interest since the observed bands in this region are very informative concerning redox state, spin state and ligation pattern of the heme iron. The high frequency RR spectra of as-isolated wild type DsrJ and DsrJC46S were therefore recorded upon Soret band excitation at 413 nm (Figure 3.16). The v_4 band is mainly a marker band for the redox state of the heme iron while v_3, v_2 and v_{10} bands are sensitive indicators of the heme spin state (Spiro, 1975; Spiro & Czernuszewicz, 1995). The RR spectra of DsrJ revealed a v_4 band centered at ~1373 cm^{-1} typical for ferric cytochrome c (Spiro & Strekas, 1974). In the low-spin ferric configuration v_3 is found above 1500 cm^{-1} and v_2 above 1579 cm^{-1}, while the v_{10} marker band is found above 1635 cm^{-1}. In the high-spin state, v_3 is located below 1500 cm^{-1}, v_2 below 1576 cm^{-1} and the v_{10} marker band is found below 1633 cm^{-1} (Rojas et al., 1997). The v_3, v_2 and v_{10} bands in the DsrJ RR spectra were observed at wave numbers of ~1505 cm^{-1}, ~1587 cm^{-1} and 1639 cm^{-1}, respectively unambiguously indicating the absence of high spin species in ferric DsrJ.

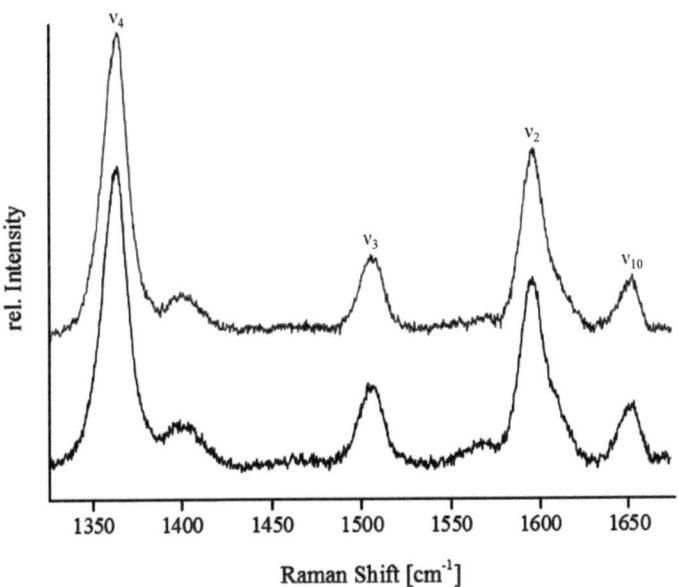

Figure 3.16: High frequency RR spectra of wild type DsrJ (red) and DsrJC46S (black) measured with 413 nm excitation and 5 mW laser power; protein concentration was 75 μM each; marker bands are labeled in the upper spectrum.

The RR spectra of wild type DsrJ and DsrJC46S were very similar (Figure 3.16) revealing that the mutation of axial Cys46 to Ser had no consequences on the RR spectra. Component analysis revealed the same band positions, line-widths and relative intensity ratios of the marker bands for the two proteins (Smilja Todorovic, personal communication). The observed modes of the spectra in Figure 3.16 appeared broadened and asymmetric. This is typically observed when heme compounds with different ligation patterns are present in the preparation. Component analysis by simulation can be used to identify the differently ligated hemes that contribute to the observed spectrum. Component analysis of the RR spectrum of DsrJ is shown in Figure 3.17. It revealed two bands for each mode of line-widths consistent with the respective natural line-widths (Siebert & Hildebrandt, 2008) which can be attributed to two heme populations (termed population A and B, respectively) with different coordination patterns (Table 3.2).

3 Results

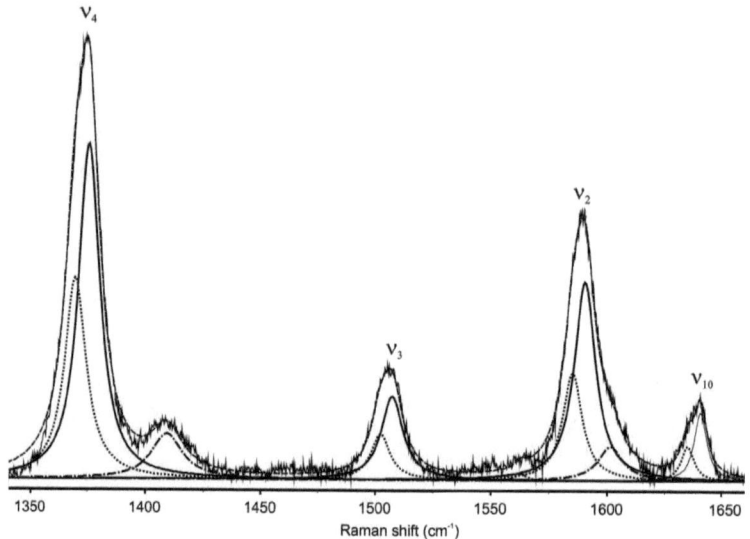

Figure 3.17: Component analysis of the high frequency RR spectrum of ferric DsrJ; experimental spectrum (thin full line), component spectra: population A (thick full line), population (dotted line) and consensus spectrum (dashed line); marker bands are labeled, the components simulated by the dashed dotted line are not sensitive to axial ligation, or redox state.

The population with up-shifted marker band frequencies (population A) bears strong spectroscopic similarities with imidazole complexes of horse heart cytochrome c or ferric cytochrome c'' (Siebert & Hildebrandt, 2008) and can therefore be assigned to the bis-His hemes.

The RR fingerprint of population B, also encountered in various cytochromes c carrying the His/Met axial ligation, such as mitochondrial cytochrome c, FixL (*Bradyrhizobium japonicum*) or cytochrome c from subunit II of caa_3 oxygen reductase (*Rhodothermus marinus*), indicates His/Met coordination (Todorovic et al., 2008). The marker band frequencies of the two populations and the assigned axial ligation pattern are listed in Table 3.2.

Table 3.2: Marker band frequencies of the two heme populations from the simulation shown in Figure 3.17 and their axial ligation assignment.

population	marker band frequencies [cm^{-1}]				axial ligation assignment
	v_4	v_3	v_2	v_{10}	
A	1375	1507	1590	1640	His/His
B	1370	1502	1584	1635	His/Met

3 Results

Since there are no reported RR spectral parameters for His/Cys coordination pattern in literature, it is not clear to which of the two spin populations it contributes. However, based on the analogy with RR spectra of low spin ferric cytochrome P450 with proximal Cys and distal H_2O/OH ligation, or substrate bound cytochrome P450 (Todorovic et al., 2006), a high similarity of the His/Cys ligated heme in DsrJ with His/Met coordinated hemes can be expected.

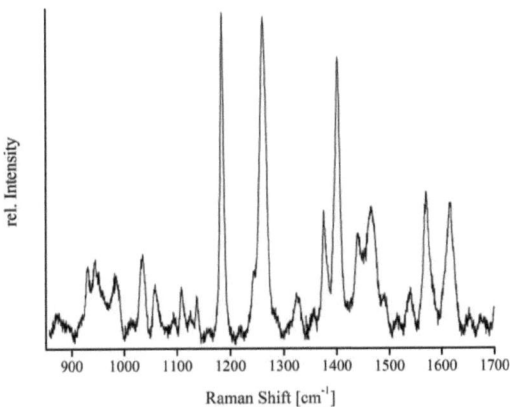

Figure 3.18: RR spectrum of wild type DsrJ measured with 647 nm excitation and 15 mW laser power; protein concentration was 75 µM.

The 650 nm band that was detected in the UV/Vis spectrum was also analysed by RR spectroscopy. Excitation of the DsrJ with 647 nm Kr laser line, in resonance with the λ_{650}nm band gave origin to a complex RR spectrum (Figure 3.18, Table 3.3).

Table 3.3: Band and mode assignment of the bands visible in the RR spectrum in Figure 3.18.

frequency [cm^{-1}]	1177	1253	1318	1394	1435	1460	1568	1615
band assignment	ν_{30}	$\nu_5 + \nu_9$	δ_{CH}[1]	ν_{20}	n.a.[2]	ν_{28}	ν_{11}	n.a.[2]
mode assignment	B_{2g}	A_{2g}		A_{2g}		B_{2g}	B_{1g}	

[1] δ_{CH} = C-H deformation of the thioether substituent

[2] n.a. = not assigned

The most intense bands in the spectrum appeared to be non-totally symmetric. Non-totally symmetric modes are due to B term scattering and can be attributed to Q bands. Therefore, the

650 nm electronic absorption band can be attributed to a Q band which is frequently associated with high spin species. However, there is no spectroscopic evidence for the presence of high spin species neither in the EPR nor RR spectra of DsrJ.

Redox titrations

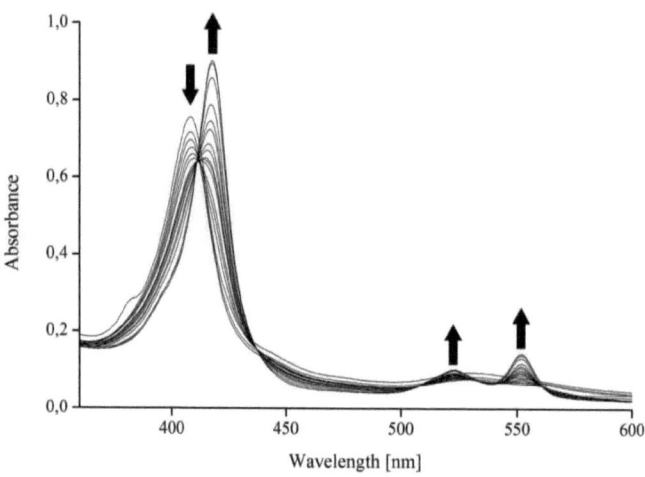

Figure 3.19: UV/Vis spectra recorded during spectrophotometric redox titration of 3 µM DsrJ; arrows indicate the absorbance changes that occurred during stepwise reduction with sodium dithionite.

Redox titrations monitoring the absorbance changes at specific wavelengths by UV/Vis spectroscopy were performed to estimate the E_m of wild type DsrJ and DsrJC46S. The absorbance changes that occurred during stepwise reduction of DsrJ can be seen in Figure 3.19.

A representative redox titration for DsrJ wt and DsrJC46S is shown in Figure 3.20. The hemes started to be reduced at ~+50 mV and are fully reduced at ~-200 mV. The experimental points showed two redox transitions with a relative proportion of 2:1, and could be fitted by adding three Nernst equations with redox potentials of -20, -200 and -220 mV in a 1:1:1 ratio. Interestingly, the redox titration of the Cys46Ser mutant protein yielded very similar results revealing that the mutation did not have any detectable effect on the redox potential of the hemes (Figure 3.20).

3 Results

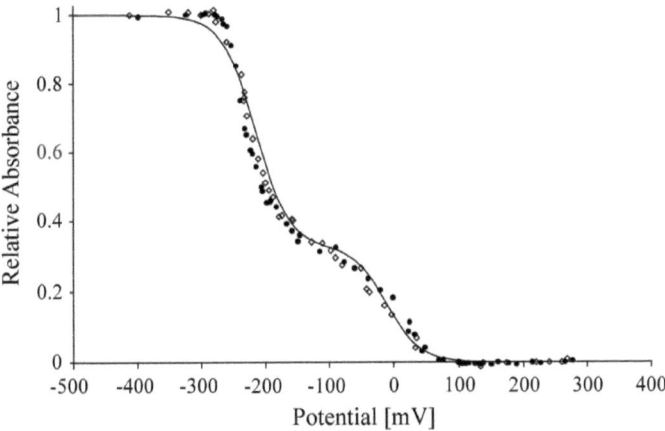

Figure 3.20: Plotted results of representative redox titrations of wild type DsrJ (full circles) and DsrJC46S (open diamonds) followed by UV/Visible spectroscopy at 417.5 nm. The full line corresponds to a theoretical simulation obtained by adding three Nernst equations with redox potentials of -20, -200 and -220 mV.

3.3 Investigating the role of cysteine 46 *in vivo*

The *A. vinosum* Δ*dsrJ* mutant strain is lacking a functional *dsrJ* gene and it has been shown that this mutant is completely unable to oxidize stored sulfur (Sander et al., 2006). To examine the role cysteine 46 of DsrJ plays *in vivo*, *A. vinosum* Δ*dsrJ* was complemented *in trans* with genes encoding either unaltered DsrJ or DsrJC46S. The constructs including the native signal peptide were cloned downstream of the *dsr* promoter region in pBBR1MCS2-L (Lübbe et al., 2006) where they replaced the *dsrL* gene. Complementation plasmids were transferred from *E. coli* S17-1 to *A. vinosum* Δ*dsrJ* by conjugation as described in section 2.7.7. Cultivated photolithoautotrophically on 1-2 mM sulfide, *A. vinosum* wild type forms sulfur globules rapidly. Under the standard conditions that are applied to study the phenotype of *A. vinosum* strains, stored sulfur is usually oxidized to the end product sulfate within 24 hours. The *A. vinosum* Δ*dsrJ* mutant complemented with unaltered *dsrJ* under the control of the *dsr* promoter oxidized sulfur within 24 hours. Accordingly sulfate was produced in a similar concentration (Figure 3.21). The phenotype of the mutant was therefore comparable to the wild type, revealing that the complementation led to a functional DsrMKJOP complex. However, when the mutant lacking *dsrJ* was complemented with the gene encoding DsrJ carrying the Cys46Ser exchange, the ability to further metabolize stored sulfur was dramatically impaired since after 24 hours only 0.16 mM sulfate was produced. Even at the end of the experiment after almost 80 hours only 0.25 mM sulfate was produced and 0.83 mM sulfur was still detected in the mutant (Figure 3.21). This finding is especially remarkable since redox potentiometric analysis had revealed similar values for all three hemes in the wild type and the mutant protein.

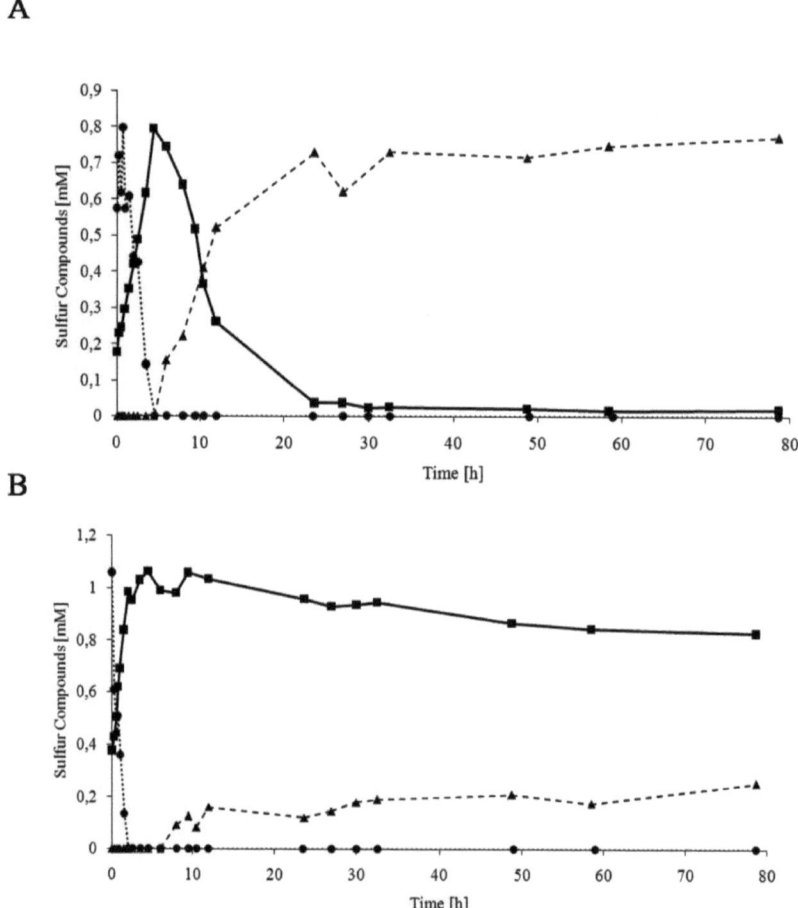

Figure 3.21: Sulfide oxidation, sulfur accumulation and oxidation of *A. vinosum ΔdsrJ* + DsrJ (A) and *A. vinosum ΔdsrJ* + DsrJC46S (B); sulfide (dotted line); elemental sulfur (full line); sulfate (dashed line); protein concentrations were 110 – 120 µg ml^{-1} at the start of the experiment.

3.4 Individual production and characterization of DsrM

3.4.1 Heterologous production of recombinant DsrM

Two distinct plasmids were used for the expression of DsrM in *E. coli*. Plasmid pMExN was constructed that encoded DsrM to which a His-tag was fused at the N-terminus. Therefore, *dsrM* was amplified from chromosomal *A. vinosum* DNA using primers M1f and M1r that introduced restriction sites for *Nde*I and *Bam*HI, respectively. Using the respective restriction enzymes, the fragment was inserted into pET-15b.

It is generally well known that the position at which a tag is fused to the protein (i.e. the C-terminus or the N-terminus) can significantly influence the production level. Additionally the position of the tag can affect proper heme incorporation when cytochromes are heterologously produced (Londer et al., 2002). It has furthermore been shown that the length of the tag reduces the production level of membrane proteins (Gordon et al., 2008; Mohanty & Wiener, 2004). In pMExN, the fusion of the His-tag to DsrM led to the addition of 20 amino acids due to a thrombin restriction site encoded in pET-15b. In order to change the location of the His-tag and to reduce the length of the tag, the vector pMexC was constructed in which a His-tag was directly fused to the C-terminus of DsrM by PCR, leading to the addition of only six amino acids. Primers M1f and Mhis2rev were used for the amplification of *dsrM*. While M1f introduced an *Nde*I restriction site, primer Mhis2rev fused a His-tag coding sequence followed by a *Bam*HI restriction site to the *dsrM* gene. *Nde*I and *Bam*HI were used to insert the DNA fragment into pET11a.

DsrM was produced from both constructs under identical conditions. The cells harvested after overexpression of *dsrM* from pMExC were significantly more colored than those expressing *dsrM* from pMExN. Next, it was clarified whether this difference was due to an unequal expression level or due to altered heme incorporation. Therefore, membrane fractions of both expression cultures were prepared in equal volumes and subjected to SDS-PAGE and subsequent Western blot analysis using anti-DsrM antibody. As seen in Figure 3.22 the amount of protein that was produced from plasmid pMExC was significantly larger than the one produced from plasmid pMExN. Expression of *dsrM* for the following analyses was therefore carried out from pMExC. 450 µg DsrM were purified from 1 liter expression culture.

3 Results

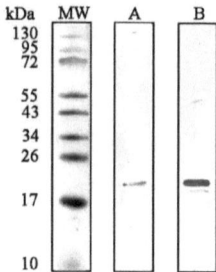

Figure 3.22: Detection of DsrM by Western blot analysis using anti-DsrM antibody; membranes were prepared from cells harbouring pMexN (A) and pMexC (B), respectively; equal fraction volumes were loaded onto the gel; (MW) prestained molecular weight marker.

E. coli C43 (DE3) was found to be best suited for the production of DsrM. For production an overnight culture was used to inoculate 500 ml of 2×YT medium that was supplied with ampicillin and the cells were grown at 37°C and 180 rpm until an OD_{600} of 0.5 was reached. IPTG was added in a final concentration of 1 mM and the medium was additionally supplied with 0.4 mM aminolevulinic acid. Cells were further grown at 25°C and 180 rpm overnight and harvested by centrifugation. Cells were washed in lysis buffer (50mM NaH_2PO_4, 300 mM NaCl) and stored at -20 °C until used.

3.4.2 Purification of DsrM

For protein purification, cells were thawed, resuspended in lysis buffer and broken by sonication. Insoluble material was removed and the membrane fraction was prepared by ultracentrifugation. The resulting pellet was resuspended in lysis buffer and 1 % (w/v) Dodecylmaltosid (DDM) was added. Solubilization was carried out with gentle stirring on ice for 1-3 hours, followed by another ultracentrifugation. This step was repeated twice. Supernatants containing solubilized protein were combined, applied to a nickel-chelate affinity chromatography and eluted with a step gradient of imidazole according to the manufacturer's instructions (Quiagen, Hilden, Germany). All buffers contained 0.1 % DDM. DsrM was purified almost to homogeneity as only minor contaminants were detected upon SDS-PAGE analysis (Figure 3.23A). As observed for the protein from *A. vinosum* (compare section 3.1.3), heterologously produced DsrM showed an anomalous migration in the Tricine-SDS-PAGE, as it ran below the 26 kDa band of the molecular weight marker (Figure 3.23). This is most probably due to the anomalous binding of SDS as commonly observed in membrane proteins (Rath et al., 2009).

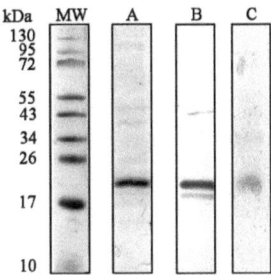

Figure 3.23: Identification of purified recombinant DsrM in Coomassie stained 10 % Tricine-SDS-PAGE (A), Western blot analysis using anti-DsrM antibody (B) and in-gel heme staining in 10 % Tricine-LDS PAGE (C); (MW), prestained molecular weight marker.

3.4.3 Characterization of DsrM

Heme content

First indication that recombinant DsrM binds heme was obtained by the red color of the cells expressing *dsrM* as compared to negative control cells harboring plasmid pET11a. Furthermore, a positive signal was observed when purified DsrM was subjected to heme staining after Tricine-LDS-PAGE (Figure 3.23C).

Pyridine hemochrome spectra of DsrM showed α and β peaks at 555 and 524 nm, respectively. The position of the peaks revealed the presence of hemes *b* (Berry & Trumpower, 1987). The heme content in DsrM was calculated to be 1.4 heme per protein. This is close to the 2 hemes *b* predicted from the sequence. Mass spectrometry was performed for final verification that proper holoprotein has been purified. DsrM was therefore subjected to LC-MS since an extensive loss of heme has been observed upon MALDI-TOF mass spectrometry of heme *b* containing proteins (Zehl & Allmaier, 2004). In contrast to MALDI-TOF, LC-MS offers a mild form of ionization that should be more suited for the analysis of cytochromes with non-covalently bound heme. However, mass spectrometry of membrane proteins is known to be difficult and the trials to clarify whether two hemes are bound to purified DsrM failed. In summary, it is obvious that the majority of protein bound heme although it cannot be excluded that some amount of protein did not fully incorporate 2 hemes *b*.

UV/Vis spectroscopy

The UV/Vis spectrum of ferric DsrM exhibited a Soret peak at 413 nm and a δ peak at 355 nm. Upon reduction, the Soret peak shifted to 426 nm and α and β peaks were observed at 559 and 530 nm, respectively. The α peak was very broad and could be better resolved at 77 K revealing a split α band with maxima at 554 of and 559 nm (Figure 3.24). Similar spectra have been observed for heterodisulfide reductases from *Methanosarcina barkeri* (Heiden et al., 1993; Heiden et al., 1994) and *Methanosarcina thermophila* (Simianu et al., 1998) and also for HmeCD purified from *Archaeoglobus profundus* (Mander et al., 2004). These enzymes are membrane bound via the subunits HdrE and HmeC, respectively, which are related to DsrM. The almost similar intensities of the two peaks in the low temperature spectra suggest that two distinct types of hemes are present in equal amounts in DsrM as has been proposed for HdrE from *M. thermophila* (Simianu et al., 1998).

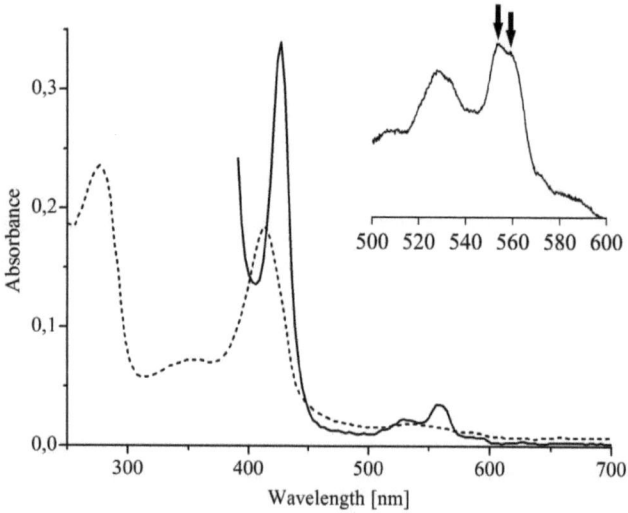

Figure 3.24: UV/Vis spectra of purified DsrM in the oxidized state (dashed line) and in the reduced state (full line); the inset is a close up of the 500 – 600 nm region recorded at 77 K; the arrows indicate the two maxima of the split α band.

EPR Spectroscopy

The EPR spectrum of *E. coli* membranes containing DsrM revealed a signal with g_{max} = 2.97 and g_{med} = 2.25 while negative control membranes revealed negligible features in this area (Figure 3.25). Other features that can be observed in both types of membranes are the g = 6.0 that can be attributed to high spin iron compounds, the g = 4.3 signal that is due to non-specific iron and other signals with g values lower than 2.

3 Results

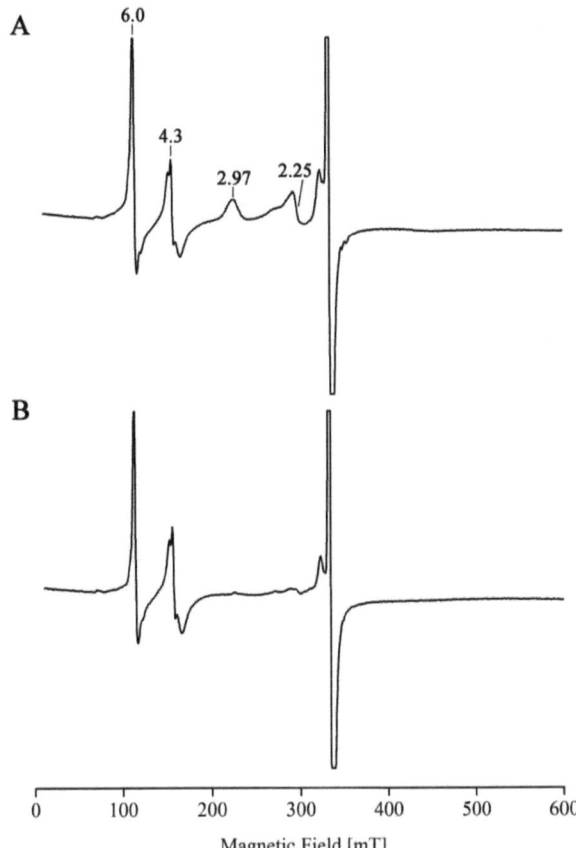

Figure 3.25: X-band EPR spectra of *E. coli* membranes expressing *dsrM* (A) and control membranes (B); Experimental conditions: microwave frequency, 9.38 GHz; microwave power, 2 mW; modulation frequency, 100 kHz; and modulation amplitude, 1 mT; temperature, 10 K; most intense signals are labeled in (A).

Since the intensities of these peaks are very similar in the control membranes, it can be ruled out that any of the signals can be attributed to a paramagnetic species in DsrM. This includes the signal for high-spin heme compounds and it can therefore be concluded that both hemes in DsrM are in low-spin state.

Reduction assays with menadiol

To investigate a putative interaction with quinones, DsrM was subjected to reduction assays using menadiol – a soluble menaquinol analogue – under anoxic conditions. The reduced minus oxidized difference spectrum indeed revealed, that DsrM was reduced after addition of menadiol as a shifting of the Soret peak was observed (Figure 3.26). Comparison with the difference spectra calculated after the addition of dithionite revealed that only a partial reduction occurred with menadiol. However, the reduction was found to be specific as adding a 50-fold molar excess of the much stronger reductant sodium borohydride had no effect on the spectrum of DsrM (not shown). These results indicate that the protein interacts with quinones *in vivo*.

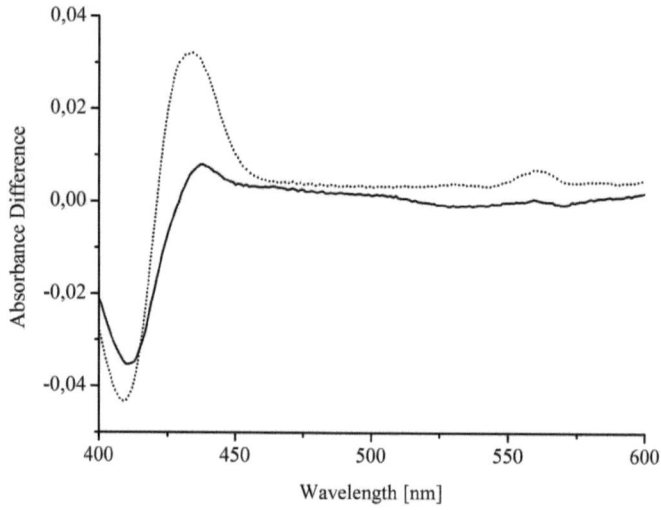

Figure 3.26: Reduced minus oxidized difference spectra of DsrM menadiol reduction assays; after addition of menadiol (solid line); after addition of dithionite (dotted line).

Redox titration

Redox titrations monitoring the absorbance changes in the visible spectrum were performed to estimate the midpoint redox potentials of the hemes in DsrM. Calculations of redox titrations with absorbance changes of the Soret peak at 426 nm and the α peak at 559 nm revealed similar results. The absorbance changes that occurred during stepwise reduction of DsrM are shown in Figure 3.27. Upon initial reduction, the Soret peak at 413 nm decreased and shifted stepwise to 426 nm. The intensity of the peak at 426 nm increased stepwise until full reduction was achieved. The intensities of broad α peak located at 559 and the β peak located at 530 nm increased upon continuative reduction. Monitoring the shape of the α peak during the redox titration revealed no major changes that would be suited for a differentiation between the two hemes that contribute to this peak.

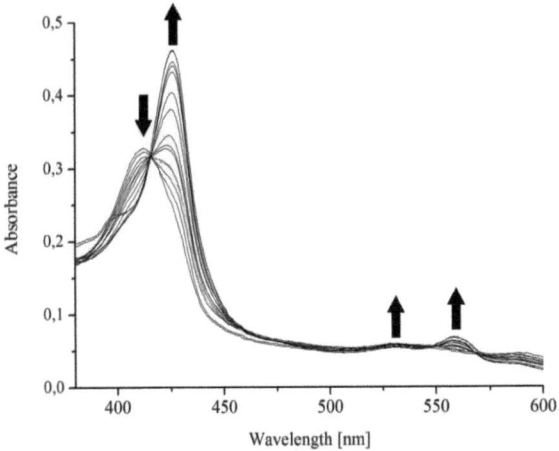

Figure 3.27: UV/Vis spectra recorded during spectrophotometric redox titration of 1.6 µM DsrM; arrows indicate the absorbance changes that occurred during stepwise reduction with sodium dithionite.

The results of a representative redox titration monitored at 426 nm are plotted in Figure 3.28. The hemes start to be reduced at ~+200 mV and are fully reduced at ~-50 mV. The experimental points are best fitted by adding two Nernst equations with redox potentials of +60 mV and +110 mV in a 1:1 ratio.

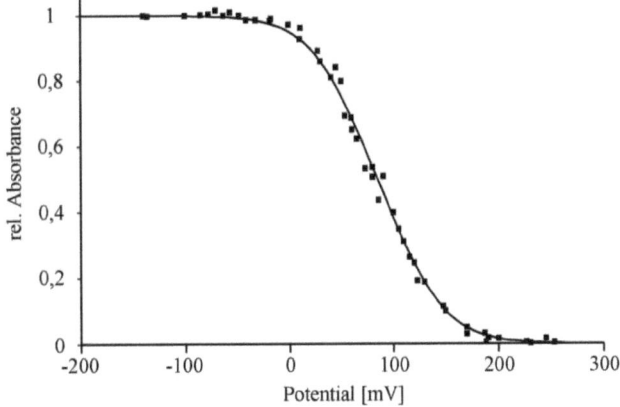

Figure 3.28: Plotted results of a representative redox titration of purified DsrM followed by UV/Visible spectroscopy at 426 nm. The full line corresponds to a theoretical simulation obtained by adding two Nernst equations with redox potentials of +60 and +110 mV.

3.5 Individual production and characterization of DsrP

3.5.1 Heterologous production of recombinant DsrP

Plasmid pPexC was constructed for the expression of *dsrP* in *E. coli*. The gene was amplified from chromosomal DNA using primers DsrPNde-f2 and DsrPEco-r and the resulting fragment was ligated into pET22b. Expression was found to be most efficient in *E. coli* C41 (DE 3). The expression conditions that were used for the production of DsrM were also found to be suitable for the production of DsrP. These conditions were therefore used with the exception, that no aminolevulinic acid was added upon induction.

3.5.2 Purification of DsrP

Purification of DsrP was carried out as described for DsrM. However, only minor amounts of protein (50 µg per liter expression culture) could be purified and the obtained sample still contained some contaminants (Figure 3.29A). In Tricine-SDS-PAGE gels, recombinant DsrP ran with an approximate molecular weight of 34 kDa which is significantly smaller than the calculated molecular weight of 46.7 kDa. As already described for DsrM, this is most likely due to anomalous migration that is commonly observed in membrane proteins (Rath et al., 2009). Due to the lack of an anti-DsrP antibody, an anti-His-tag antibody was used to identify the purified protein (Figure 3.29B). Two bands were observed in each case, which most likely represent the monomeric and the dimeric form of the protein.

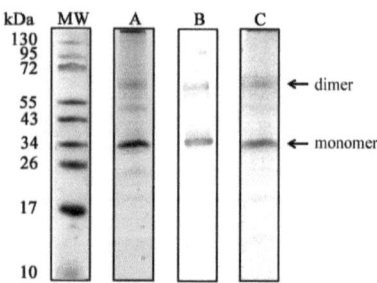

Figure 3.29: Identification of purified recombinant DsrP in Coomassie stained 10 % Tricine-SDS-PAGE (A), Western blot analysis using anti-His-tag antibody (B) and in-gel heme staining in 10 % Tricine-LDS PAGE (C); (MW), prestained molecular weight marker.

3.5.3 Characterization of DsrP

Heme binding of DsrP

Despite the fact that no hemes were expected from sequence analysis, the protein that eluted from the Strep-Tactin column surprisingly revealed a reddish color. Additionally, a positive signal in heme stained Tricine-LDS-PAGE gels could be observed (Figure 3.29C) for monomeric and dimeric DsrP (Figure 3.29C). This revealed that DsrP binds heme and ruled out that the minor contaminants were responsible for the color of the preparation.

UV/Vis spectroscopy

Final identification of DsrP as a *b*-type cytochrome was achieved by UV/Vis spectroscopy. The spectrum revealed that the protein as isolated was partly reduced as evident by a shoulder at the Soret peak and by the fact that α- and β peaks were already slightly visible (not shown). Ferricyanide was used to oxidize the cytochrome and the Soret peak was observed at 412 nm. After addition of dithionite, the Soret peak shifted to 425 nm and α- and β peaks were observed at 559 and 528 nm, respectively (Figure 3.30). UV/Vis redox titrations were not performed due to the low yield of recombinant protein.

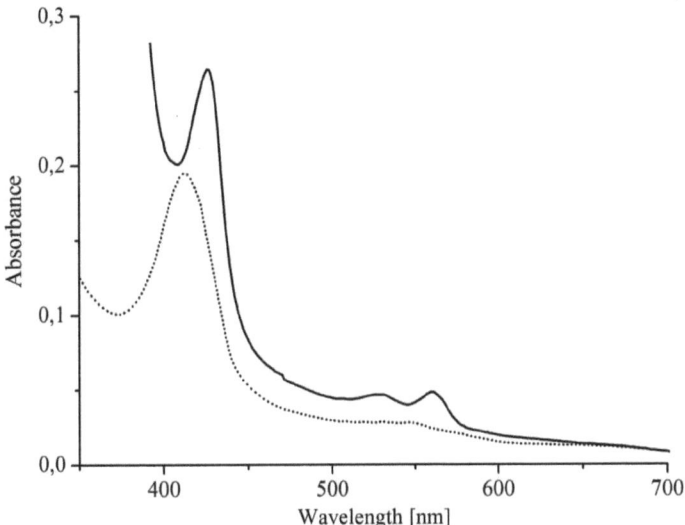

Figure 3.30: UV/Vis spectra of purified DsrP in the oxidized state (dashed line) and in the reduced state (full line).

Reduction assays with menadiol

Due to the similarity of DsrP to other quinone active membrane subunits, it was tested whether DsrP can be reduced with menadiol or not. The difference spectrum clearly showed that the Soret peak shifted to 428 nm after addition of menadiol and an increased absorbance in the α band region was observed, yielding evidence for a partial reduction of the cytochrome (Figure 3.31). This reduction was found to be specific since no reduction occurred when sodium borohydride was added to the solution in a 50-fold molar excess.

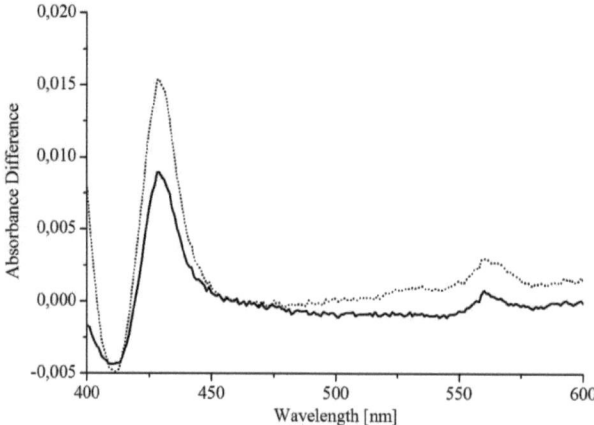

Figure 3.31: Reduced minus oxidized difference spectra of DsrP menadiol reduction assays; after addition of menadiol (solid line); after addition of dithionite (dotted line).

Bioinformatic analysis of DsrP

As already mentioned, DsrP was so far not thought to bind heme b. This assumption was based on bioinformatic analyses where no highly conserved putative ligands were identified (Dahl et al., 2005; Pires et al., 2006) and furthermore on the fact that heme binding has never been experimentally proven for any protein of the NrfD protein family. With the knowledge that DsrP does bind heme b, bioinformatic analyses were repeated focusing on the identification of putative heme ligands.

3 Results

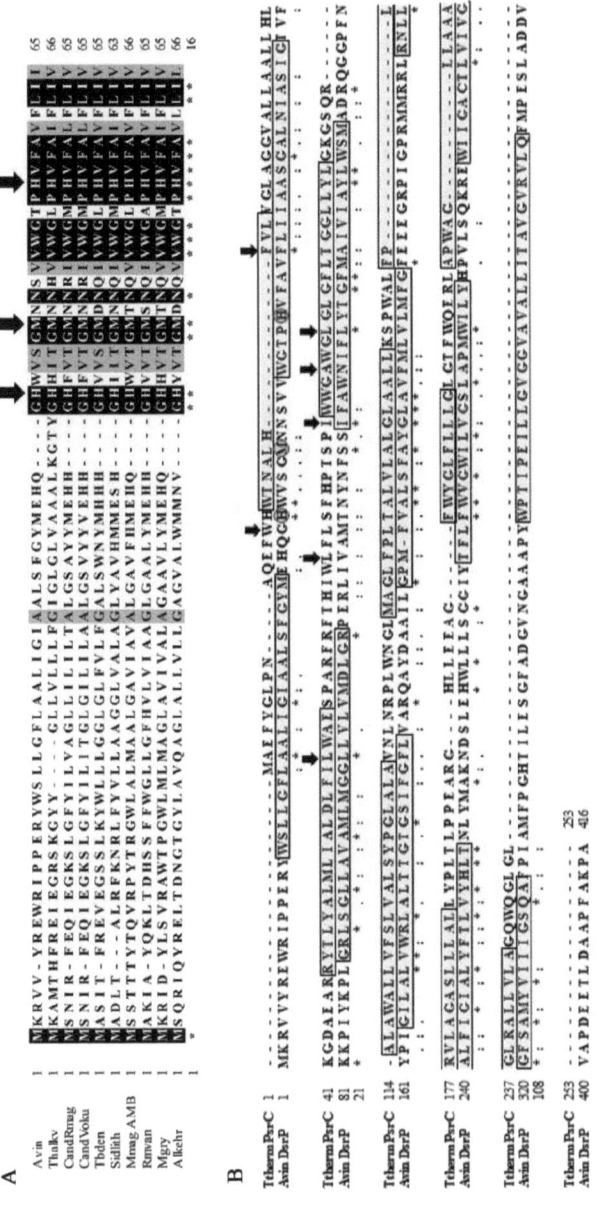

Figure 3.32: Sequence alignment of the N-terminus of DsrP proteins from proteobacterial sulfur oxidizers; identical amino acids are boxed in black, similar amino acids are boxed in grey, conserved amino acids that could be involved in heme ligation are marked with an arrow; Avin, *Allochromatium vinosum*; Thalkv, *Thioalkalivibrio* sp. HL-EbGR7; CandRmag, *Candidatus* Ruthia magnifica str. Cm; CandVoku, *Candidatus* Vesicomyosocius okutanii*; Tbden, *Thiobacillus denitrificans*; Sidlith, *Sideroxydans lithotrophicus*; Mmag AMB, *Magnetospirillum magnetotacticum* AMB; Rmvan, *Rhodomicrobium vanilii*; Mgry, *Magnetospirillum gryphiswaldense*; Alkehr, *Alkalilimnicola ehrlichii*. (B), Sequence alignment of *Thermus thermophilus* PsrC (Ttherm PsrC) and *Allochromatium vinosum* DsrP (Avin DsrP); transmembrane helices are

3 Results

Sequence analysis revealed three conserved amino acids that may serve as axial heme ligands: Histidine 42, methionine 47 and histidine 57 (*A. vinosum* numbering) Figure 3.32A). Especially the histidines are likely candidates, since they are located in or close to transmembrane helices (Figure 3.32B). These conserved residues are only present in proteobacterial sulfur-oxidizing bacteria, and they are absent in sulfate-reducing bacteria and in the *Chlorobiaceae* (Pires et al., 2006).

Recently the crystal structure of PsrC from *Thermus thermophilus* has been solved and the amino acids that are involved in quinone binding were identified (Jormakka et al., 2008). A sequence alignment between PsrC and DsrP was performed since DsrP is closely related to PsrC from *Thermus thermophilus* (Figure 3.32B). The amino acids that are involved in quinone binding in PsrC are marked with an arrow and the alignment shows that the majority of these residues is conserved in DsrP. When these conserved residues are drawn in a topology model of DsrP it can clearly be seen that the putative quinone binding site, which may be formed amongst others by those amino acids, is located on the periplasmic side of the membrane (Figure 3.33). Additionally it becomes obvious that the putative quinone binding site is close to the proposed axial heme ligands that are located in or close to the first two transmembrane helices.

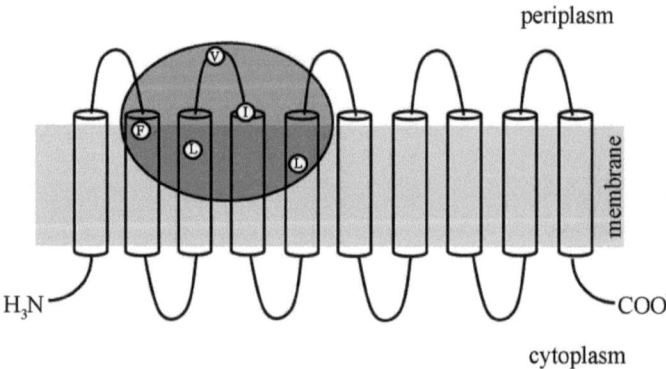

Figure 3.33: Topology model of DsrP, the conserved amino acids that may be involved in quinone binding are plotted.

3.6 Individual production and characterization of DsrK

3.6.1 Heterologous production of recombinant DsrK and cellular localization

Plasmid pPRIBAdsrK was constructed for the expression of *dsrK*. The plasmid encoded DsrK including a Strep-tag sequence fused to the C-terminus. Using primers DsrKEcoRI-f and DsrKEco47III-r, *dsrK* was amplified from chromosomal DNA and restriction sites for *EcoRI* and *Eco47III* were introduced at the 5' and the 3' end of the gene, respectively. Using these restriction enzymes *dsrK* was ligated into pPR-IBA 1 resulting in plasmid pPRIBAdsrK. In most cases expression of *dsrK* led to the formation of inclusion bodies that sedimented upon low speed centrifugation ($< 10000 \times g$). Inclusion bodies consist of miss- or unfolded insoluble protein and their formation is frequently observed upon recombinant protein production in heterologous hosts. However, when *E. coli* C41 (DE 3) was used as expression host, some amount of protein was found to be correctly folded when the following conditions were applied: An overnight culture was used to inoculate 400 ml of LB medium supplied with ampicillin. The cells were grown at 37°C and 180 rpm until an OD_{600} of 0.6 was reached. Expression was induced by the addition of 0.1 mM IPTG. After two hours under the same conditions the cells were harvested by centrifugation.

To determine the cellular localization of the recombinant protein cells were lysed by sonication and the membrane fraction was separated from the soluble fraction by ultracentrifugation. The membranes were resuspended in the same volume of the soluble fraction and the samples were applied to SDS-PAGE followed by western blot analysis using an anti-DsrK antibody. As shown in Figure 3.34 DsrK was detected as ~60 kDa protein and it was exclusively found in the membrane fraction. Proteins that are loosely associated with the membrane can often be washed off the membranes by using high ionic strength or by changing the pH of the buffer. However under all conditions applied, DsrK was always exclusively found in the membrane fraction. Therefore, Triton X-100 was used to solubilize the membranes and indeed DsrK could be detected in the supernatant after solubilization (Figure 3.34).

3 Results

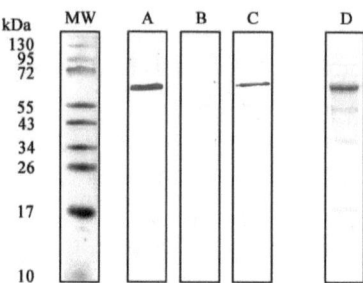

Figure 3.34: Identification of recombinant DsrK via Western blot analysis using anti-DsrK antibody; in the *E. coli* membrane fraction (A), in the soluble fraction (B) and in the supernatant after solublization (C); purified DsrK in Coomassie stained Tricine-SDS-PAGE (D); (MW) prestained molecular weight marker

3.6.2 Purification of DsrK

Cells were resuspended in buffer W (100 mM Tris-HCl, 150 mM NaCl, pH 7.5) and broken by sonication. Preparation of the membrane fraction and solubilization was carried out as described for DsrM (see section 3.4.2) with the exception that Triton X-100 was used as detergent. DsrK was purified by Strep-Tactin affinity chromatography according to the manufacturer's instructions (IBA Tagnology, Göttingen, Germany). All buffers contained 0.1 % Triton X-100. DsrK was successfully purified as shown in Figure 3.34D although only small amounts were obtained (200 µg per liter expression culture) due to the low expression of properly folded protein.

3.6.3 Characterization of DsrK

In vitro reconstitution of FeS clusters

From sequence analysis, DsrK is proposed to bind three FeS clusters. When purified from *E. coli* the protein contained only minor amounts of FeS clusters as often observed when FeS proteins are expressed in *E. coli*. This became obvious as the preparation revealed only minor absorbance in the 300 - 500 nm region in the UV/Vis spectrum (Figure 3.35). Therefore, *in vitro* reconstitution of the FeS clusters was accomplished under anoxic conditions. After *in vitro* reconstitution the solution revealed a brownish color and the UV/Vis spectrum indicates the incorporation of FeS clusters (Figure 3.35) as the spectrum showed a broad absorbance between 300 and 500 nm. However the amount of protein was insufficient for analysis of the FeS clusters by EPR spectroscopy. Therefore, further characterization of the FeS clusters was not carried out.

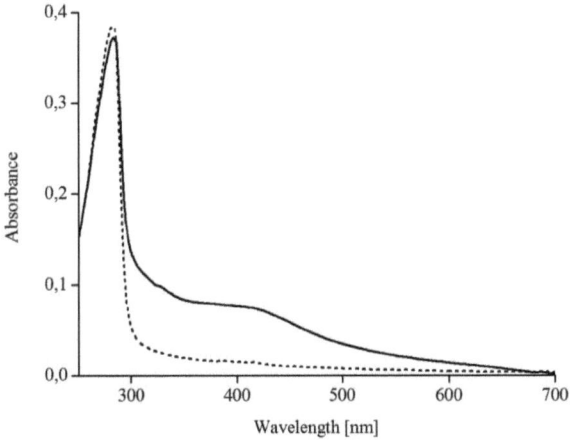

Figure 3.35: UV/Visible spectra of purified recombinant DsrK as prepared (dotted line) and after *in vitro* reconstitution of the FeS clusters (full line).

Membrane binding of DsrK in *A. vinosum*

It has previously been reported that DsrK is associated with the membrane fraction of *A. vinosum* since it can be purified from the membrane fraction and it has been suggested that DsrM or DsrP serve as membrane anchors for the protein (Dahl et al., 2005). However, this was not experimentally verified and the results with the recombinant protein indicated, that it integrates into membranes on its own. To clarify this point, membrane fractions of *A. vinosum* wild type and of strains carrying in frame deletions in *dsrM* and *dsrP* (Sander et al., 2006), respectively, were prepared and analyzed for the presence of DsrK by Western blot analysis after Tricine-SDS-PAGE. The rationale behind this approach was the assumption that DsrK would become a soluble protein or become unstable upon loss of its native membrane anchor. However, DsrK was clearly present not only in membranes isolated from *A. vinosum* wt but also in membranes of the Δ*dsrM* and Δ*dsrP* strains (Figure 3.36). Furthermore DsrK was neither detected in the soluble fraction of the wild type nor in those of the mutants. It can therefore confidently be ruled out that DsrK is attached to the membrane via DsrM or DsrP.

3 Results

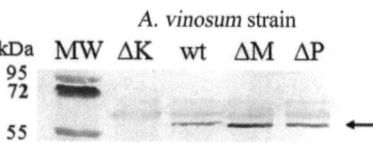

Figure 3.36: Western Blot analysis of *A. vinosum* wt and deletion mutants membrane fractions using anti-DsrK antibody; MW, molecular weight marker, ΔK, *ΔdsrK* mutant, wt, wild type; ΔM, *ΔdsrM* mutant; ΔP, *ΔdsrP* mutant; the arrow indicates the signal that corresponds to DsrK.

Membrane anchoring of DsrK

The experimental results indicated that DsrK is bound to the membrane on its own. However, no transmembrane helices are predicted for DsrK and the hydropathy profile indicates that it is mainly hydrophilic. Therefore, sequence analyses were carried out with a focus on amphiphatic α-helices. The amphiphatic character of an α-helices enables it to anchor a protein in-plane into the membrane. To identify amphiphatic α-helices, the secondary structure was predicted and the predicted α helices were analyzed using the helical wheel projection applet. The results are displayed in Figure 3.37 and they reveal the presence of an amphiphatic α-helix at the N-terminus of DsrK predicted to consist of the amino acids 50-75 (Figure 3.37A). Since the α helix consists of 26 amino acids, nine projections were made totally. The first projection (amino acids 50-68) revealed a hydrophobic and a hydrophilic part and is shown in Figure 3.37B as a representative. All other projections also revealed a hydrophobic and a hydrophilic part of the helix and their relative position in the projections indicates that the hydrophobic part is in-plane over the entire helix. Therefore, the respective helix is a likely candidate for monotopic membrane anchoring of the protein.

3 Results

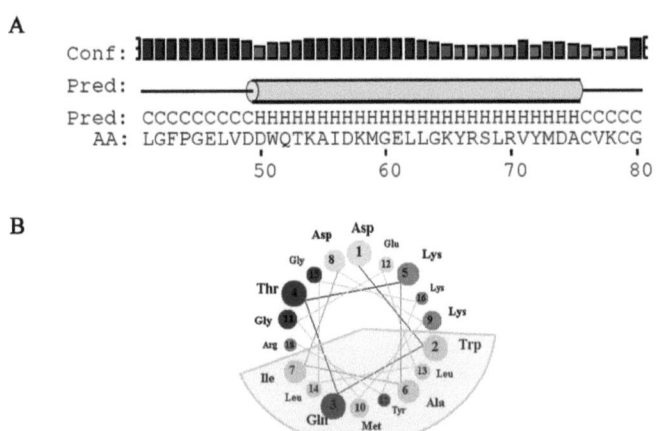

Figure 3.37: (A), predicted secondary structure of DsrK (amino acids 41-80); (Conf), confidence of prediction, the higher the bar, the more confident; (Pred), prediction; graphical (top), the tube symbolizes the α-helix; letter code (below), C = coil, H = helix; (AA), amino acid sequence (B), helical wheel projection of amino acids 50-68 of DsrK; the hydrophobic part of the helix is shaded.

Additionally the online program AMPHIPASEEK was used for the detection of amphiphatic α helices in DsrK. This program offers two modi for the identification of in-plane membrane anchors (Sapay et al., 2006). In modus 1 the propability for an in-plane membrane anchor is predicted from the amino acid sequence of the protein. In modus 2 an alignment is calculated first and the prediction is carried out independently for all sequences in the alignment. Finally a consensus prediction is made. In both modi at least some of the amino acids in the α helix shown in Figure 3.37 were predicted to form an in-plane membrane anchor.

Interaction of DsrK with DsrC

In order to detect a putative interaction between DsrK and DsrC, the interaction of recombinant DsrK and DsrC was analyzed by coelution assays taking advantage of the Strep-tag fused to the C-terminus of DsrK. After incubation, a mixture of DsrK and DsrC was applied to a Strep-Tactin superflow column (IBA BioTagnology, Göttingen, Germany) followed by purification according to the manufacturer's instructions. A control experiment was done under the same conditions with the exception that DsrK was omitted in sample that was applied to the Strep-Tactin column. This control experiment revealed as expected, that DsrC does not bind to the affinity matrix as it eluted during the first wash steps (Figure 3.38A). However, when DsrC was incubated with DsrK carrying a Strep-tag before being applied to the Strep-Tactin matrix, hardly any DsrC was detected in the washing fractions and the majority of DsrC eluted specifically together with DsrK (Figure 3.38B).

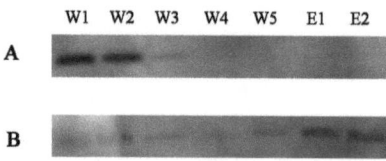

Figure 3.38: DsrK – DsrC coelution assays; detection of DsrC using anti-DsrC antibody; (A), negative control without DsrK; (B), DsrC plus DsrK; W1-W5, washing fractions 1 to 5; E1-E2, elution fractions 1 and 2.

3.7 Individual production and initial characterization of DsrO

3.7.1 Heterologous production of recombinant DsrO

DsrO contains an N-terminal signal peptide for the translocation into the periplasm via the Tat pathway. The proteins that are transported via this pathway are folded inside of the cytoplasm and cofactors are inserted prior to protein export into the periplasm (Berks et al., 2005). Protein translocation and processing of the protein is often a bottleneck during heterologous production of proteins (Li et al., 2004). DsrO was therefore produced as a cytoplasmic protein carrying an N-terminal Strep-tag instead of the signal peptide. By using primers DsrOEcorRI-f2 and DsrOXho-r, *dsrO* was amplified and the resulting fragment was restricted with *Eco*RI and *Xho*I. The fragment was then ligated into plasmid pASK-IBA5 that had been restricted with the same restriction enzymes.

For the production of DsrO, an overnight culture of *E. coli* BL21 (DE3) harboring IBAdsrO was used to inoculate 400 ml LB medium supplied with ampicillin. Cells were grown until an OD_{600} of 0.6 was reached and expression was induced by the addition of anhydrotetracycline in a final concentration of 0.2 µg ml^{-1}. After 3 hours the cells were harvested by centrifugation. To improve the content of FeS clusters in DsrO, the medium was additionally supplied with 0.5 mM ammonium ferric citrate and 0.5 mM cysteine. In another experiment the *E. coli isc* genes were coexpressed from plasmid pACYCisc (Gräwert et al., 2004) which additionally encodes a resistance for chloramphenicol that was accordingly added to the medium.

3.7.2 Purification of DsrO

Cells were resuspended in buffer W and lysed by sonication. After removing insoluble cell material by centrifugation, DsrO was purified by Strep-Tactin affinity chromatography according to the manufacturer's instructions (IBA Tagnology, Göttingen, Germany). The protein that was successfully purified as seen by SDS-PAGE analysis was identified as DsrO by Western blot analysis using a specific anti-DsrO antibody (Figure 3.39). 3.12 mg DsrO were obtained from 1 liter expression culture.

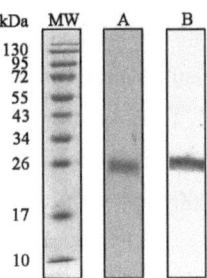

Figure 3.39: Identification of purified recombinant DsrO; (A), Coomassie stained 15 % SDS-PAGE; (B), Western blot analysis using anti-DsrO antibody; (MW), prestained molecular weight marker.

3.7.3 Initial characterization of DsrO

UV/Vis spectroscopy and chemical analysis of DsrO

When DsrO was purified from *E. coli*, the resulting protein solution revealed a brownish color and the UV/Vis spectrum indicated the presence of FeS clusters (Figure 3.40). The spectrum shows maxima at 325 and 417 nm with a shoulder at 460 nm.

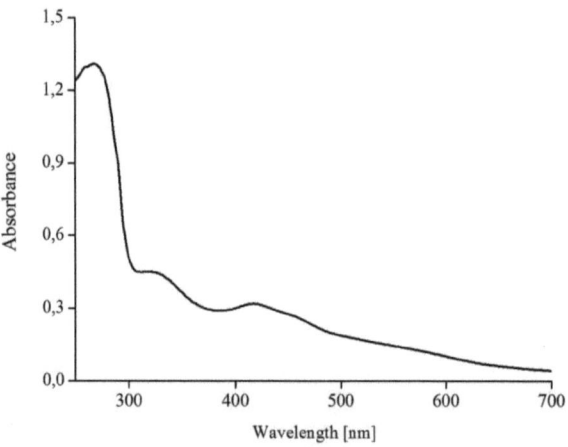

Figure 3.40: UV/Vis spectrum of purified DsrO in the as-isolated state.

Upon addition of dithionite, the overall absorbance decreased without a change in the position of the observed peaks (not shown). However, these spectroscopic properties indicated the presence of [2Fe-2S] clusters rather than the presence of the expected [4Fe-4S] clusters (Lippard, 1995). A very similar spectrum was observed for IscA that binds a [2Fe-2S] cluster (Wollenberg et al., 2003).

The amount of iron and sulfur ions in DsrO was further investigated by the quantification of iron and sulfur. The quantification revealed only 4.0 (± 0.1) mol iron and 3.9 (± 0.2) mol sulfur per mol protein instead of the 16 mol that were expected for a protein that binds four [4Fe-4S] clusters. Varying different parameters of the expression conditions such as expression medium, temperature or time and strength of induction had no effects. Furthermore adding ammonium ferric citrate and cysteine to the medium and coexpressing the *E. coli isc* genes from pACYCisc (Gräwert et al., 2004) did not change the spectroscopic properties of DsrO and had negligible positive effects on the iron and sulfur content of the recombinant protein. Additionally, purification carried out under anoxic conditions revealed the same results ruling out that a cluster breakdown and conversion from [4Fe-4S] to [2Fe-2S] clusters had occurred. Such a reaction has been reported for the FNR protein from *E. coli* upon exposure to oxygen (Khoroshilova et al., 1997).

4 DISCUSSION

The main aim of the present work was the characterization of the individual components of the DsrMKJOP transmembrane complex. Therefore, each of the proteins was heterologously produced in *E. coli* followed by its purification and characterization. Notably, the production of a protein in absence of its interaction partners can have a significant influence on the proteins properties as it has for instance been reported for the *E. coli* nitrate reductase (Magalon et al., 1997; Rothery et al., 1999). In order to verify the results obtained for the proteins individually produced in *E. coli*, DsrMKJOP was also enriched from A. *vinosum*. First trials for purification of DsrMKJOP from *A. vinosum* membranes have been carried out via classical chromatograpical methods following the enrichment of DsrK (Dahl et al., 2005). However, the complex obtained appeared to be unstable under the applied conditions since no cytochromes were detected by UV/Vis spectroscopy or heme staining. While substoichiometric amounts of DsrJ could be detected by the use of a specific antibody, the presence of DsrM and DsrP in the preparation could not be verified (Dahl et al., 2005). Instead, large amounts of dissimilatory sulfite reductase (dSiR) were co-purified. The presence of this siroheme-containing protein prevented further spectroscopic characterization of the membrane-associated Dsr components. In this work, a different strategy was applied and *dsrJ* that was engineered with a His-tag-coding sequence was expressed in a $\Delta dsrJ$ *A. vinosum* host. Thus, the DsrMKJOP complex was successfully enriched by nickel-chelate affinity chromatography without co-purification of dSiR. Notably, DsrM was positively identified in Western blot analyses using an anti-DsrM antibody, whereas it has so far never been detected in *A. vinosum* (Sander, 2005).

In order to study the individual components of the DsrMKJOP transmembrane complex, each of the proteins was heterologously produced in *E. coli*. In the case of DsrJ, preliminary work had already been done that however did not yield satisfying production of holoprotein (Schneider, 2007; Dahl et al., 2005). In the present work, DsrJ was successfully produced as a soluble protein, transported into the periplasm, processed and properly maturated to a triheme cytochrome *c*. Minimal expression without induction with IPTG and the coexpression of the *E. coli ccm* genes from plasmid pEC86 (Arslan et al., 1998) were found to be important factors for correct maturation. It was shown that the native signal peptide of DsrJ is not

4 Discussion

cleaved off when it is produced in *E. coli* and that it anchors the protein into the membrane. The apparent molecular mass of native DsrJ from *A. vinosum* as judged by SDS-PAGE and comparison with the soluble version from *E. coli* after removal of the PelB leader allows the conclusion that the signal peptide of DsrJ is not cleaved off in *A. vinosum*. A similar retainment of the signal peptide has been shown for the analogous protein from *Archaeoglobus fulgidus* (Mander et al., 2002). It has been proposed that the signal peptide serves as a membrane anchor for this protein which can accordingly also be proposed for DsrJ in *A. vinosum*.

DsrJ was studied by UV/Vis, EPR and RR spectroscopies, as well as by mass spectrometry and redox potentiometry. The existence of an unusual His/Cys heme axial coordination was shown to be present and its role was investigated by site-directed mutagenesis followed by biochemical and biophysical characterization of the protein. These studies were further complemented by *in vivo* experiments.

For the identification and characterization of axial heme ligands in cytochromes, point mutations are typically introduced to change the putative native ligands to other well established heme ligands like histidine or methionine. The midpoint redox potentials of hemes mainly depend on their molecular environment and on the axial ligation of the iron (Paoli et al., 2002; Wallace & Clark-Lewis, 1992; Battistuzzi et al., 2002). Therefore, a ligand exchange usually leads to an alteration of the midpoint redox potentials, depending on the introduced mutation (Alric et al., 2004; Aubert et al., 2001; Barker et al., 1996). Compared to histidine/cysteine or bis-histidine ligated hemes, the midpoint potentials of His/Met ligated hemes are typically higher (Paoli et al., 2002; Reedy et al., 2008). Another strategy is replacement of the natural ligand by small residues like alanine or glycine that cannot ligate the heme iron (Nakajima et al., 2001; Ran et al., 2007). In this study, a different strategy was used to identify and examine the potential role of a cysteine as the sixth axial ligand of one of the three hemes in DsrJ. Cysteine 46 was here replaced with serine. These two amino acids only differ by the presence of an oxygen atom in serine instead of the sulfur atom in cysteine. The missing signal at g_{max}= 2.517 in the DsrJC46S EPR spectrum, typical for thiolate-coordinated hemes unambiguously revealed that cysteine 46 may coordinate one of the hemes in DsrJ. The cysteine coordination is however most likely only partial as shown by spectral integrations of the EPR spectra and by the fact that the typical signal for His/Cys ligated heme decreases upon reduction and re-oxidation. The absence of a high-spin heme indicates that the

4 Discussion

cysteine is most likely replaced by another residue. A second conserved methionine that is present in DsrJ (Figure 1.2) is a likely candidate. This is presumably also the case in DsrJC46S, although in theory the introduced serine could also act as the sixth axial ligand. Serine has not been reported as a heme ligand in biological systems yet (Reedy & Gibney, 2004). Therefore, spectroscopic data for comparison are not available. Nevertheless, heme ligation by homoserine in a semisynthesised peptide led to a six-coordinated high spin heme (Wallace & Clark-Lewis, 1992) which is apparently not present in DsrJC46S.

The UV/Vis spectrum of DsrJ reveals a band at 650 nm in the oxidized and in the reduced state. Resonance Raman spectra revealed only the presence of non-totally symmetric modes, indicating that the 650 nm electronic absorption band belongs to a Q band (α/β transitions). Similar bands are frequently present in the spectra of ferric P450 cytochromes in which Cys is a proximal ligand (Yoshioka et al., 2001). Some authors have correlated this band with five-coordinated high-spin species, but it is also present in some substrate bound LS P450 cytochromes, and also in some Cys/His P450 mutants (Martinis et al., 1996). The presence of high spin hemes in DsrJ is ruled out by both EPR and RR spectroscopy, but the fact that the 650 nm band is also present in the C46S mutant also argues against it being due to Cys coordination. Thus, it is not clear at this point which heme gives origin to this band

Redox titrations revealed the presence of three redox species in DsrJ with midpoint potentials of -20 mV, -200 mV and -220 mV, respectively. This simplified interpretation of the results most likely does not reflect the actual redox potential of the three hemes, as several confounding factors are likely present: First, the differently coordinated hemes may have distinct extinction coefficients and will therefore contribute differently to the absorbance at any given wavelength; second, it is likely that there is redox cooperativity between the hemes, as commonly observed in multi-heme cytochromes (Pereira & Xavier, 2005), which affects the macroscopic redox potentials determined in the redox titration. In particular, there seems to be cooperativity in the two hemes being reduced around -210 mV, suggesting a close contact between them. Since the proposed His and Met axial ligands are adjacent in the sequence (Figure 1.2), the hemes that they coordinate will most likely be in close proximity, and are good candidates for this interacting heme pair. Interestingly, the redox titration of the Cys46Ser mutant protein yielded very similar results revealing that the mutation did not have any detectable effect on the redox potential of the hemes. This further indicates, as suggested from the EPR results, that only a small fraction of one heme may be coordinated by His/Cys,

4 Discussion

and this fraction may decrease with reduction. Furthermore, these results suggest that a possible His/Cys coordinated heme is not likely to be involved in a redox function. Nevertheless, is can confidently be stated that the hemes in DsrJ start to be reduced at ~+50 mV and are fully reduced at ~-200 mV. This is in contrast to the His/Cys ligated heme present in the purified DsrMKJOP complex from *D. desulfuricans* which was still not fully reduced at -400 mV (Pires et al., 2006). While the midpoint potential of the His/Cys coordinated heme in PufC from *R. sulfidophilum* was determined to be -160 ±10 mV (Alric et al., 2004), the potentials of His/Cys ligated hemes in SoxXA from *Chlorobaculum* (formerly *Chlorobium*) *tepidum, Paracoccus pantotrophus* and *Starkeya novella* are all below -400 mV or are even redox inactive in some cases (Kappler et al., 2008; Ogawa et al., 2008; Reijerse et al., 2007). For *A. vinosum* DsrJ a very negative midpoint potential or redox inactivity of the His/Cys ligated heme can confidently be excluded. Rather the heme already started to be reduced upon addition of the mild reductant ascorbate and could be fully reduced with dithionite as shown by EPR analysis. However, care must been taken in extrapolating redox potentiometric analysis of a single component to the *in vivo* situation within the complex. Therefore, an *A. vinosum* membrane fraction that was enriched in DsrMKJOP was investigated as well. The redox behavior of the hemes *c* in this preparation was very similar to that of isolated recombinant DsrJ.

The role and significance of His/Cys ligation in *c*-type cytochromes is not understood yet. It has been most studied in SoxXA proteins, which are involved in thiosulfate oxidation (Friedrich et al., 2005; Kelly et al., 1997). The catalytic subunit SoxA is a diheme protein in *R. sulfidophilum* and *P. pantotrophus* (Cheesman et al., 2001; Rother & Friedrich, 2002) while heme 1 is missing in *S. novella* and *C. tepidum* (Kappler et al., 2004; Ogawa et al., 2008). In all cases, heme 2 or the single heme corresponding to heme 2 has His/Cys ligation and is characterized by a very negative midpoint potential (Kappler et al., 2008; Ogawa et al., 2008; Reijerse et al., 2007). When present, heme 1 is also His/Cys ligated and redox inactive (Cheesman et al., 2001; Reijerse et al., 2007). In SoxXA from *P. pantotrophus*, heme 1 remains oxidized even at potentials of -800 mV (Reijerse et al., 2007) and structural investigations of the proteins from *R. sulfidophilum* and *P. pantotrophus* revealed that the distance between heme 1 and heme 2 is too big for electron transfer (Bamford et al., 2002; Dambe et al., 2005). Redox inactivity and the fact that heme 1 is replaced by a disulfide bridge in the protein from *S. novella* (Kappler et al., 2004) lead to the conclusion that this

4 Discussion

His/Cys ligated heme is not important for the catalytic function of SoxXA (Reijerse et al., 2007). Heme 2 has an exceptionally low midpoint potential of below -400 mV in all SoxXA proteins studied so far and is characterized by the modification of cysteine to a persulfide that ligates the heme (Bamford et al., 2002; Cheesman et al., 2001). This heme is the catalytic active site. The substrate thiosulfate is bound to the ligating cysteine resulting in a SoxA-thiocysteine-S-sulfate intermediate. The sulfur-sulfur bond is polarized by an arginine and the thiosulfate may be bound to SoxYZ (Bamford et al., 2002). It is interesting to note that there is a conserved arginine in close vicinity to cysteine 46 in DsrJ (Figure 1.2). So far, there is no evidence for a persulfide modification of the cysteine in DsrJ.

It was observed that the His/Cys coordination for one of the hemes is most likely only partial, and that the Cys46Ser mutation has almost no effect on the redox behavior of the protein *in vitro*, while *in vivo* this ligand exchange led to a protein that was functionally strongly impaired. Thus, it appears rather unlikely that the dramatic effect of ligand exchange *in vivo* would solely be due to interrupted electron transport. Furthermore, heme ligand switching has notably never been reported for a heme protein that is "only" involved in electron transfer. Instead, ligand switching has been reported for heme sensor proteins such as CooA and *Bx*RcoM-2 (Marvin et al., 2008; Yamashita et al., 2004) and - more interestingly - for the cytochrome cd_1 nitrite reductase from *Paracoccus pantotrophus*. The latter enzyme catalyzes the one-electron reduction of nitrite to nitric oxide and is has been shown that a switch in heme ligation is required for enzymatic activity (Allen et al., 2000; Cheesman et al., 1997; van Wonderen et al., 2007).

The results presented in this work are consistent with a scenario where one heme may be present in two different coordination states, namely His/Cys and His/Met, with this change in coordination being related to a possible catalytic function of Cys46 in sulfur chemistry. In this context it has to be noted, that there are two DsrJ proteins in the database where cysteine 46 (*A. vinosum* numbering) is lacking and replaced by a histidine: The proteins from the endosymbionts of *Calyptogena magnifica* (Figure 1.2) and *Calyptogena okutanii*. Furthermore, the two methionines, that are likely candidates for the alternative heme ligation or the ligation of another heme, are also missing (Figure 1.2). However, no experimental data concerning the Dsr proteins are available for these organisms.

4 Discussion

DsrM was successfully purified from *E. coli* C43 (DE3) membranes as a *b*-type cytochrome. The UV/Vis spectrum of DsrM is similar to spectra observed for heterodisulfide reductases from *Methanosarcina barkeri* (Heiden et al., 1993; Heiden et al., 1994) and *Methanosarcina thermophila* (Simianu et al., 1998) and also for purified HmeCD from *Archaeoglobus profundus* (Mander et al., 2004). These enzymes are membrane bound via the subunits HdrE and HmeC, respectively that are related to DsrM. In contrast to the UV/Vis spectra, the EPR spectrum of DsrM differs from those of HdrE and HmeC. One of the hemes in the latter proteins is in high spin state (Mander et al., 2004; Simianu et al., 1998) while both hemes in DsrM were found to be in low spin state. This is also consistent with the DsrMKJOP preparation from *D. desulfuricans* where no high spin heme was observed (Pires et al., 2006). HdrE, HmeC and DsrM belong to the nitrate reductase gamma subunit superfamily (cl00959) the most prominent member of which is NarI from *E. coli*. The nitrate reductase NarGHI from *E. coli* is a well characterized enzyme that catalyzes the oxidation of quinole by nitrate that results in the generation of a proton gradient across the membrane (see Blasco et al. (2001) and Rothery et al. (2008) for reviews). NarI and the properties of the hemes have been extensively studied (Rothery et al., 2001). Like DsrM, the protein contains five transmembrane helices and binds two hemes *b* where each heme is axially ligated by one histidine from helix II and helix V, respectively. One heme is located towards the periplasmic side of the membrane bilayer and is referred to as the distal heme (heme b_D), whereas the other is located towards the cytoplasmic side and is referred to as the proximal heme (heme b_P). It has been shown that the proximal heme has a relatively low redox potential of +20 mV and the distal heme has a relatively high redox potential of +120 mV (Rothery et al., 1999). Therefore, the two hemes are also often referred to as heme b_L and heme b_H, respectively (Magalon et al., 1997; Rothery et al., 1999). The hemes function as a conduit for electron flow from a quinone binding site, that has been identified on the periplasmic side of the NarI in close vicinity to heme$_D$ and the electrons are then transferred via the FeS clusters in NarH to the catalytic subunit NarG (Bertero et al., 2005; Lanciano et al., 2007; Magalon et al., 1997; Rothery et al., 2001). The localization of the hemes in DsrM is most likely similar to NarI since the histidines that serve as axial heme ligands are conserved in DsrM and located within the same transmembrane helices (Pott & Dahl, 1998). The crystal structure of NarGHI has been solved and the residues in the quinone binding site were identified (Bertero et al., 2005). The quinone binding site in DsrM might be slightly different since none of these residues is

4 Discussion

conserved in DsrM. This may be due to the fact that menaquinol instead of ubiquinol could be the substrate for DsrM since *A. vinosum* contains both types of quinoles (Imhoff, 1984). However, the fact that DsrM can be specifically reduced with menadiol strongly suggests that the protein interacts with the quinone pool. With the determined midpoint potentials of +60 and +110 mV for the hemes an electron flow similar to the situation in NarI, namely from the quinole via heme b_L and b_H to the adjacent subunit can be proposed for DsrM. Similar electron flow has also been proposed for HmeC from *Archaeoglobus* species and for HdrE from *Methanosarcina* species (Mander et al., 2004; Mander et al., 2002; Heiden et al., 1994). The electron donor of the latter is the more negative methanophenazine and in accordance to that, the midpoint redox potentials of the hemes in *M. thermophila* HdrE were determined to be -23 and -180 mV (Simianu et al., 1998).

DsrP was successfully purified from *E. coli* C41 (DE 3) and positively identified in Western blot analysis using an anti-His-tag antibody as well as in a Coomassie stained Tricine-SDS-gel and in heme staining. The detection of the protein in Tricine-SDS-PAGE is quite remarkable, since the presence of related subunits in several CISM holoenzymes has been inferred from genetic experiments or nucleotide sequence analysis but corresponding bands were not visible upon denaturing polyacrylamide gel electrophoresis due to their highly hydrophobic character (Dietrich & Klimmek, 2002; Hussain et al., 1994; Krafft et al., 1995; Laska et al., 2003; Schröder et al., 1988). Indeed, DsrP and its homologue HmeB were not detected in the preparations from *D. desulfuricans* and *Archaeoglobus fulgidus*, respectively (Mander et al., 2002; Pires et al., 2006).

DsrP was clearly identified as *b*-type cytochrome by UV/Vis spectroscopy and heme staining. This is highly remarkable since it was so far believed, that DsrP does not contain heme (Dahl et al., 2005; Pires et al., 2006). Heme binding has only been debated for the related HybB, the membrane subunit of hydrogenase 2 from *E. coli* (Menon et al., 1994; Dubini et al., 2002), whereas the first structure of a protein from this family, PsrC from *T. thermophilus* clearly shows that it does not contain heme (Jormakka et al., 2008). This suggests that some of the members of the NrfD protein family may contain heme *b* and some may not. It should be noted that DsrP provides the first experimental evidence for heme binding in this family. Sequence analysis revealed the presence of three amino acids that are conserved within proteobacterial sulfur-oxidizing bacteria and which may serve as axial heme ligands. These conserved residues are absent in sulfate-reducing bacteria and in the sulfur-oxidizing

Chlorobiaceae (Pires et al., 2006). This is in line with the fact that the Dsr proteins of the *Chlorobiaceae* are more closely related to the proteins from sulfate reducers than to those from *A. vinosum* (Sander et al., 2006). Since no hemes *b* were detected in DsrP from *D. desulfuricans* (Pires et al., 2006), it is possible that only DsrP proteins from proteobacterial sulfur-oxidizing bacteria bind heme.

Reduction with menadiol suggests that DsrP interacts with the quinone pool and the topology model of DsrP together with the similarity to PsrC strongly suggests a quinone binding site on the periplasmic side of the protein. Although the real distance depends certainly on the three-dimensional structure of the protein, the putative quinone binding site is most likely close to the conserved amino acids that may serve as axial heme ligands for the heme *b*, as deduced from the sequence. This is expected for a quinone active cytochrome as the distance between redox centers within proteins is usually below 14 Å (Page et al., 2003). While DsrM may work as quinol oxidase donating electrons to DsrK, DsrP could act as quinone reductase, just as the related HybB. DsrM and DsrP could thus be connected via the quinones. Although quinones are known to mediate electron (and proton) transport over relatively long distances within the membrane, quinones are also involved in intramolecular electron transport. For instance, quinones shuttle between different redox centers within an intramolecular cavity in the cytochrome b_6f complex from cyanobacteria (see Cramer & Zang (2006) and Cramer et al. (2006) for reviews).

The fact that DsrK was found in the membrane fraction of *E. coli* together with the finding that neither DsrM nor DsrP are the membrane anchors for DsrK in *A. vinosum* identifies DsrK as membrane protein. Since the protein is mainly hydrophilic and no transmembrane helices are predicted for DsrK, alternative membrane anchoring is most likely. Monotopic membrane proteins are bound to the membrane interface. The membrane anchor can be made of covalent links to a hydrophobic compound, electrostatic binding to phospholipid head groups, hydrophobic loops inserted into the membrane interface or amphiphatic α-helices inserted into the membrane interface (Sapay et al., 2006). These helices are referred to as in-plane membrane anchors. Monotopic membrane anchoring via an in-plane membrane anchor can be proposed for DsrK since an amphiphatic α helix was detected. A similar membrane anchoring has been reported for SdhE (formerly SdhC) from *Acidianus ambivalens* that is another member of the CCG domain superfamily (Lemos et al., 2001). Although some amount of recombinant DsrK could be purified from *E. coli* membranes and *in vitro* reconstitution

4 Discussion

revealed the presence of FeS clusters, the total protein yield was too low for EPR spectroscopy. Therefore, it was not possible to elucidate if a $[4Fe-4S]^{3+}$ cluster - expected to be ligated by the CCG domain – is present in DsrK. Such a $[4Fe-4S]^{3+}$ cluster has been detected in heterodisulfide reductase (HDR) upon addition of HS-CoM and it is believed to be an intermediate during enzymatic turnover of HDR (Hedderich et al., 2005). A similar paramagnetic species has been observed in DsrMJKOP from *D. desulfuricans* and in the Hme complex from *Archaeoglobus fulgidus*, however in the as-isolated forms without addition of a putative substrate (Pires et al., 2006; Mander et al., 2002). At present it is uncertain if this is due to a tightly bound thiol substrate in the active sites of the purified proteins or if it is due to a difference in the catalytic site relative to HDR (Pires et al., 2006). DsrK contains only one of the CCG domains, while the members of the CCG domain superfamily contain two. Nevertheless, it has more recently been shown that only the C-terminal domain – the one that is conserved in DsrK - binds the catalytic FeS cluster whereas the other domain provides ligands to a Zn site (Hamann et al., 2009; Hamann et al., 2007). The similarity between DsrK and the catalytic subunit of HDR has led to the suggestion that a thiol-disulfide interchange reaction is catalyzed by DsrK and DsrC has been suggested to be its substrate (Dahl et al., 2005; Pires et al., 2006). DsrC is encoded in the same operon as the DsrKMJOP complex (Dahl et al., 2005). The protein has two strictly conserved cysteine residues at its Carboxyterminus and these have been shown to form specific disulfide bridges (Cort et al., 2008) that could be acted upon by DsrK. In support of this hypothesis, the coelution assays clearly show that DsrK and DsrC interact with each other. The interaction of DsrC with the DsrMKJOP complex is currently also investigated in *D. vulgaris* and first unpublished results support this result (I. A. C. Pereira, Sofia S. Venceslau, personal communication).

Expression of *dsrO* and subsequent purification of the protein resulted in a brownish protein which UV/Vis spectrum indicated the presence of FeS clusters. The UV/Vis spectrum and the iron and sulfur quantifications however revealed that most probably [2Fe-2S] clusters were bound in DsrO. The DsrMKJOP complex from *D. desulfuricans* and the Hme complex from *Archaeoglobus fulgidus* do not contain [2Fe-2S] clusters suggesting that the wrong cluster form has been inserted by the heterologous host. Varying the expression and purification conditions had almost no effect on the UV/Vis spectra and/or the iron and sulfur contents. The insertion of wrong cluster forms has also been reported for other FeS proteins that were produced in *E. coli* (Hamann et al., 2009; Hamann et al., 2007).

4 Discussion

Proposed function of the DsrMKJOP transmembrane complex. The structure of the *D. vulgaris* dissimilatory sulfite reductase has been solved with DsrC bound in a cleft between DsrA and DsrB. A mechanism for the process of sulfite reduction including DsrAB, DsrC and DsrMKJOP has been proposed in which DsrAB persulfurates DsrC at one of its two conserved cysteine residues (Oliveira et al., 2008). DsrC then dissociates from the sulfite reductase followed by the reduction of the persulfide by the second conserved cysteine. This results in release of H_2S and the formation of an intramolecular disulfide in DsrC. For regeneration the disulfide may be reduced by the DsrMKJOP transmembrane complex (Oliveira et al., 2008). The two redox active cysteines in DsrC also play a central role in the model by Cort et al. (2008) for sulfur oxidation in *A. vinosum*. In this model DsrC is involved in sulfur transfer reactions together with DsrEFH, a protein that is found in sulfur-oxidizing but not in sulfate-reducing organisms. DsrC is suggested to act as a substrate donating protein for sulfite reductase. In sulfur oxidizers this enzyme is thought to operate in the opposite direction as compared to sulfate reducers (Schedel et al., 1979). In one of the suggested models for sulfur oxidation in *A. vinosum* sulfide is reductively released as a substrate for sulfite reductase and an intramolecular disulfide is proposed to be formed in DsrC similar to the situation suggested for *Desulfovibrio* species (Cort et al., 2008). This disulfide has to be reduced to regenerate free thiol groups to restart the cycle. The coelution assays presented here have shown that DsrK indeed interacts with DsrC and the reduction of the DsrC disulfide by DsrK can be proposed in close analogy to the model for *Desulfovibrio*.

Although the DsrMKJOP complex from *D. desulfuricans* has been purified and characterized, it is not clear whether DsrJ is an electron entry or exit point, or a catalytic subunit (Pires et al., 2006). However, a more recent model for the function of the complex proposes electron transfer from the periplasm to the cytoplasm (Oliveira et al., 2008). Since the sulfur metabolism of *D. vulgaris* and *A. vinosum* are the reverse of each other, one could assume that electron flow through the DsrMKJOP complex in *A. vinosum* proceeds from the cytoplasm into the periplasm. This could account for the difference in the heme redox potentials observed between DsrJ from the two organisms. However, recently *A. vinosum* complementation experiments with *dsrJ* from *D. vulgaris* have unambiguously shown that the *dsrJ* gene from the sulfate reducer *D. vulgaris* can complement the *A. vinosum* Δ*dsrJ* mutant, restoring the wild type phenotype (Grein et al., 2010). This experiment proved that despite the

4 Discussion

fact that *A. vinosum* DsrJ shares only 34 % similarity the *D. vulgaris* DsrJ, the latter is able to substitute its homologue in *A. vinosum*, indicating that both can perform the same function in organisms with opposite sulfur metabolisms, i.e. sulfate reduction versus sulfur oxidation (Grein et al., 2010). Interesting in this respect is, as already mentioned, the pylogenetic distribution of the DsrMKJOP proteins: The proteins from the sulfur-oxidizing *Chlorobiaceae* are more closely related to the proteins from sulfate reducers than to *A. vinosum* and horizontal gene transfer from the sulfate reducers to the *Chlorobiaceae* has been proposed (Sander et al., 2006). DsrJ is therefore proposed to be the electron entry point of the complex, either directly by oxidizing a putative substrate or by accepting electrons from an interaction partner.

The putative catalytic activity of DsrJ raises the question of a possible substrate. There are several sulfur compounds that have to be considered in this context. Sulfide, Polysulfides and Thiosulfate are sulfur compounds that are known to be oxidized by *A. vinosum* and most likely their oxidation takes place in the periplasm (Frigaard & Dahl, 2009). For thiosulfate it is known that the oxidation is accomplished by the proteins encoded by the *sox* genes (Hensen et al., 2006). Furthermore, the oxidation of thiosulfate is not impaired in the *A. vinosum* Δ*dsrJ* mutant strain (Sander et al., 2006), ruling out that DsrJ is involved in the oxidation of thiosulfate. The sulfide:quinone-oxidoreductase is currently the most likely candidate for the oxidation of sulfide although this has not been finally proven (Frigaard & Dahl, 2009). Polysulfides are known to occur as intermediates upon oxidation of sulfide to sulfur (Prange et al., 2004) and accordingly *A. vinosum* can use external polysulfide as photosynthetic electron donors. The enzymes for the oxidation of polysulfides are currently not known although there are several polysulfide reductase like genes located in the genome of *A. vinosum* (C. Dahl, personal communication). Despite the fact that is has not completely been solved which proteins are responsible for the oxidation of sulfide and polysulfide, the oxidation of these compounds is unambiguously not impaired in the *A. vinosum* mutant strain that lacks a functional DsrJ (Sander, 2005; Sander et al., 2006). In contrast to this, the oxidation of sulfur globules is strictly dependent on DsrJ (Sander et al., 2006) and its cysteine46 as shown in this study. The observation that the *A. vinosum* Δ*dsrJ* and Δ*dsrJ*+DsrJC46S strains are unable to oxidize stored sulfur does however not directly indicate a catalytic activity of DsrJ since the deletion of genes encoding other (cytoplasmic) Dsr proteins led to the same phenotype (Dahl et al., 2008; Lübbe et al., 2006). The finding of a putatively catalytic activity of a periplasmic protein is quite interesting since the sulfur globules are located within the same cell compartment. Although it is still unknown how sulfur is transported into the cytoplasm for further oxidation, a reductive activation to a perthiol that is subsequently transported into the cytoplasm has been proposed (Dahl et al., 2005). However, the results of this study rather indicate the uptake of electrons by DsrJ, which is the reverse process. Furthermore, sulfur globules are not present in sulfate-reducing prokaryotes so the role of DsrJ in these organisms would still not be clear. For these reasons, it is difficult to conceive the role of DsrJ in the activation of sulfur globules.

Further sulfur compounds that have been identified in *A. vinosum* growing on sulfide are monosulfane sulfonic acids (Franz et al., 2009). Two different types of monosulfane sulfonic acids were identified. Depending on their different retention time during HPLC analysis, the compounds were termed RT15 and RT30, respectively. The compounds were exclusively found in the cells and not in the medium. Moreover, the concentration of the compounds (in particular the concentration of compound RT15) interestingly showed a tight correlation with the concentration of intracellular sulfur. When sulfur began to be oxidized, the concentration of monosulfane sulfonic acids reached its maximum and it has been proposed that these compounds may be relevant intermediates in the oxidation of stored sulfur (Franz et al., 2009). Monosulfane sulfonic acids are inorganic acids, with the general formula HS_n-SO_3H. In this context it is interesting to note that the substrate of SoxXA with its His/Cys ligated heme is thiosulfate and that thiosulfate represents the monosulfane sulfonic acid with the shortest chain length. However, the assignment of a substrate to DsrJ at this point is speculative and further research is needed to verify the catalytic activity of DsrJ and to identify its substrate.

It was shown that DsrM is a quinone reactive diheme cytochrome *b* and that the hemes have midpoint potentials of 60 and 110 mV, respectively. The significant similarity to NarI and HdrE suggests electron flow to DsrK (related to HdrD). Moreover, since in DsrMKJOP only the hemes *b* but not the hemes *c* are reducible by menadiol, electron flow from DsrM or DsrP to DsrJ – possibly via DsrO – is most unlikely. The same has also been observed for the purified Dsr complex from *D. desulfuricans* (Pires et al., 2006). This is furthermore in agreement with the difference in the determined midpoint redox potentials of the recombinant *A. vinosum* proteins as the potentials of the hemes in DsrJ are significantly more negative than those of for the hemes in DsrM.

4 Discussion

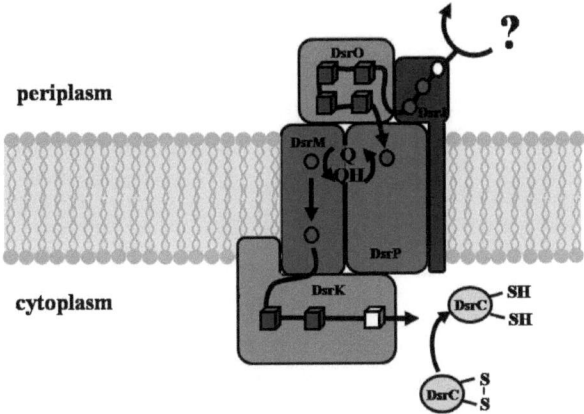

Figure 4.1: Proposed model for the function of the DsrMKJOP transmembrane complex; the questionmark represents a putative sulfur substrate or an alternative electron donor, hemes and FeS clusters are indicated with circles and cubes, respectively; possibly catalytic prosthetic groups are depicted in white.

Figure 4.1 summarizes the proposed model for the function of the DsrMKJOP transmembrane complex. DsrJ may be involved in the oxidation of a putative sulfur substrate in the periplasm or accept electrons from a yet unknown donor. The electrons would be transported across the membrane via the other components of the DsrMKJOP complex. The fact that DsrO is a periplasmic protein and the similarity of DsrOP to electron transporting subunits of a number of respiratory enzyme complexes suggests that electrons may be further transported via these proteins. Since DsrP and DsrM are both quinone interacting proteins these two proteins could be connected via quinones. Alternatively - or additionally - the heme *b* that was found in DsrP, could be involved in electron transfer from DsrP to DsrM. DsrM would then donate electrons to DsrK – the (or another) catalytic subunit of the complex. This would reduce the disulfide formed by the conserved cysteines in DsrC to generate free thiols. This would enable DsrC to restart a cycle of sulfur transfer from DsrEFH to DsrAB as suggested by Cort et al. (2008).

5 Outlook

Despite the fact that the present work provides significant insights into the function of the DsrMKJOP complex *in vivo*, several open questions remain that make the transmembrane complex to an interesting objective of further research.

Although it has been shown that electrons are most likely transported from the periplasm into the cytoplasm, the origin of the electrons is still unknown. It has been shown that DsrJ may be a catalytic subunit of the complex and further verification of this proposal is a challenging task. This includes the identification of the putative substrate and elucidation of the catalytic mechanism that may be connected with a ligand switching as in the cytochrome cd_1 nitrite reductase from *Paracoccus pantotrophus* (Allen et al., 2000; Cheesman et al., 1997; van Wonderen et al., 2007). The heme ligation in DsrJ may be further investigated by monitoring the pH dependence of redox titrations, by the sequential mutation of the His and Met axial ligands and by EPR monitored redox titrations with accurate spin counting. Furthermore, the elucidation of the three-dimensional structure of DsrJ would be an important step to understand the physiological role of this unusual cytochrome.

Further studies will also be needed to assess the electron flow within the complex. In particular the connection between the membrane spanning subunits DsrP and DsrM and the role of quinones within this connection is highly interesting. It may also be investigated if this electron transfer is coupled to proton translocation.

The heme containing proteins were in the focus of the present work. An in-depth characterization of the FeS cluster containing proteins DsrO and DsrK is a remaining task that inevitably requires detailed EPR spectroscopic analyses.

Finally the proposed catalytic FeS cluster in DsrK needs to be identified and characterized. Its role in catalytic turnover - most likely with DsrC as substrate - may be addressed.

6 SUMMARY

In the present study the DsrMKJOP transmembrane complex was investigated and characterized by biochemical, biophysical and functional analyses. Each of the proteins was successfully produced in *E. coli* as recombinant proteins and purified. It has been shown that DsrJ is a triheme cytochrome and EPR spectroscopy provided evidence for a possible, but only partial, His/Cys heme ligation in one of the hemes. This heme shows heterogeneous coordination with Met being another candidate ligand. Cysteine 46 was replaced by serine using site-directed mutagenesis, with the mutant protein showing a significant decrease in the EPR signal attributed to His/Cys coordination, but identical UV-Vis and resonance Raman spectra. The redox potentials of the hemes in the wild type protein were determined to be -20, -200 and -220 mV and were found to be virtually identical in the mutant protein. However, *in vivo* the same ligand exchange led to a dramatically altered phenotype highlighting the importance of Cys46. The results suggest that Cys46 may be involved in catalytic sulfur chemistry rather than electron transfer. It was furthermore shown that the signal peptide of DsrJ is not cleaved off in *A. vinosum* and that it may anchor the protein into the membrane. DsrM was identified as a diheme cytochrome *b* and the two hemes were found to be in low spin state. Their midpoint redox potentials were determined to be +60 and +110 mV. Although no hemes were predicted for DsrP, it was also clearly identified as a *b*-type cytochrome. As obvious from the literature, this is the first time that heme binding has been experimentally proven for a member of the NrfD protein family. Both cytochromes were partly reduced after addition of a menaquinol analogue suggesting interaction with quinones *in vivo*. DsrO and DsrK were both experimentally proven to be Fe-S-containing proteins. In addition, DsrK was shown to be membrane-associated most likely via an in-plane membrane anchor. Coelution assays provide support for the proposed interaction of DsrK with the soluble cytoplasmic protein DsrC, which might be its substrate. The results obtained in this study were combined in a model for the function of the DsrMKJOP complex in *A. vinosum*.

7 REFERENCE LIST

Alderton, W. K., Cooper, C. E., and Knowles, R. G. (2001) Nitric oxide synthases: structure, function and inhibition, *Biochem. J.* **357**, 593-615.

Allen, J. W., Watmough, N. J., and Ferguson, S. J. (2000) A switch in heme axial ligation prepares *Paracoccus pantotrophus* cytochrome cd_1 for catalysis, *Nat. Struct. Biol.* **7**, 885-888.

Alric, J., Tsukatani, Y., Yoshida, M., Matsuura, K., Shimada, K., Hienerwadel, R., Schoepp-Cothenet, B., Nitschke, W., Nagashima, K. V. P., and Verméglio, A. (2004) Structural and functional characterization of the unusual triheme cytochrome bound to the reaction center of *Rhodovulum sulfidophilum*, *J. Biol. Chem.* **279**, 26090-26097.

Altschul, S. F., Gish, W., Miller, W., Myers, E. W., and Lipman, D. J. (1990) Basic local alignment search tool, *J. Mol. Biol.* **215**, 403-410.

Arslan, E., Schulz, H., Zufferey, R., Künzler, P., and Thöny-Meyer, L. (1998) Overproduction of the *Bradyrhizobium japonicum* c-type cytochrome subunits of the cbb_3 oxidase in *Escherichia coli*, *Biochem. Biophys. Res. Commun.* **251**, 744-747.

Aubert, C., Guerlesquin, F., Bianco, P., Leroy, G., Tron, P., Stetter, K.-O., and Bruschi, M. (2001) Cytochromes c_{555} from the hyperthermophilic bacterium *Aquifex aeolicus*. 2. heterologous production of soluble cytochrome c_{555}^s and investigation of the role of methionine residues, *Biochemistry* **40**, 13690-13698.

Bamford, V. A., Bruno, S., Rasmussen, T., Appia-Ayme, C., Cheesman, M. R., Berks, B. C., and Hemmings, A. M. (2002) Structural basis for the oxidation of thiosulfate by a sulfur cycle enzyme, *EMBO J.* **21**, 5599-5610.

Barker, P. D., Nerou, E. P., Cheesman, M. R., Thomson, A. J., de Oliveira, P., and Hill, H. A. O. (1996) Bis-methionine ligation to heme iron in mutants of cytochrome b_{562}. 1. Spectroscopic and electrochemical characterization of the electronic properties, *Biochemistry* **35**, 13618-13626.

Battistuzzi, G., Borsari, M., Cowan, J. A., Ranieri, A., and Sola, M. (2002) Control of cytochrome *c* redox potential: axial ligation and protein environment effects, *J. Am. Chem. Soc.* **124**, 5315-5324.

Bazaral, M., and Helinski, D. R. (1968) Circular DNA forms of colicinogenic factors E1, E2 and E3 from *Escherichia coli*, *J. Mol. Biol.* **36**, 185-194.

Beinert, H. (2000) Iron-sulfur proteins: ancient structures, still full of surprises, *J Biol Inorg Chem* **5**, 2-15.

Beinert, H., Holm, R. H., and Munck, E. (1997) Iron-sulfur clusters: nature's modular, multipurpose structures, *Science* **277**, 653-659.

Bendtsen, J. D., Nielsen, H., von Heijne, G., and Brunak, S. (2004) Improved prediction of signal peptides: SignalP 3.0, *J. Mol. Biol.* **340**, 783-795.

Berks, B. C., Palmer, T., and Sargent, F. (2005) Protein targeting by the bacterial twin-arginine translocation (Tat) pathway, *Curr. Opin. Microbiol.* **8**, 174-181.

Berry, E. A., and Trumpower, B. L. (1987) Simultaneous determination of hemes *a*, *b*, and *c* from pyridine hemochrome spectra, *Anal. Biochem.* **161**, 1-15.

Bertero, M. G., Rothery, R. A., Boroumand, N., Palak, M., Blasco, F., Ginet, N., Weiner, J. H., and Strynadka, N. C. J. (2005) Structural and biochemical characterization of a quinol binding site of *Escherichia coli* nitrate reductase A, *J. Biol. Chem.* **280**, 14836-14843.

Blasco, F., Guigliarelli, B., Magalon, A., Asso, M., Giordano, G., and Rothery, R. A. (2001) The coordination and function of the redox centres of the membrane-bound nitrate reductases, *Cell. Mol. Life Sci.* **58**, 179-193.

Blattner, F. R., Williams, B. G., Blechl, A. E., Denniston-Thompson, K., Faber, H. E., Furlong, L., Grunwald, D. J., Kiefer, D. O., Moore, D. D., Schumm, J. W., Sheldon, E. L., and Smithies, O. (1977) Charon phages: safer derivatives of bacteriophage lambda for DNA cloning, *Science* **196**, 161-169.

Bradford, M. M. (1976) A rapid and sensitive method for the quantitation of microgram quantities of protein utilizing the principle of protein-dye binding, *Anal. Biochem.* **72**, 248-254.

Brune, D. C. (1989) Sulfur oxidation by phototrophic bacteria, *Biochim. Biophys. Acta* **975**, 189-221.

Bryantseva, I., Gorlenko, V. M., Kompantseva, E. I., Imhoff, J. F., Suling, J., and Mityushina, L. (1999) *Thiorhodospira sibirica* gen. nov., sp. nov., a new alkaliphilic purple sulfur bacterium from a Siberian soda lake, *Int. J. Syst. Bacteriol.* **49** 697-703.

Cheesman, M. R., Ferguson, S. J., Moir, J. W., Richardson, D. J., Zumft, W. G., and Thomson, A. J. (1997) Two enzymes with a common function but different heme ligands in the forms as isolated. Optical and magnetic properties of the heme groups in the oxidized forms of nitrite reductase, cytochrome cd_1, from *Pseudomonas stutzeri* and *Thiosphaera pantotropha*, *Biochemistry* **36**, 16267-16276.

Cheesman, M. R., Kadir, F. H., al-Basseet, J., al-Massad, F., Farrar, J., Greenwood, C., Thomson, A. J., and Moore, G. R. (1992) E.p.r. and magnetic circular dichroism spectroscopic characterization of bacterioferritin from *Pseudomonas aeruginosa* and *Azotobacter vinelandii*, *Biochem. J.* **286**, 361-367.

7 Reference List

Cheesman, M. R., Little, P. J., and Berks, B. C. (2001) Novel heme ligation in a *c*-type cytochrome involved in thiosulfate oxidation: EPR and MCD of SoxAX from *Rhodovulum sulfidophilum*, *Biochemistry* **40**, 10562-10569.

Cort, J. R., Mariappan, S. V., Kim, C. Y., Park, M. S., Peat, T. S., Waldo, G. S., Terwilliger, T. C., and Kennedy, M. A. (2001) Solution structure of *Pyrobaculum aerophilum* DsrC, an archaeal homologue of the gamma subunit of dissimilatory sulfite reductase, *Eur. J. Biochem.* **268**, 5842-5850.

Cort, J. R., Selan, U., Schulte, A., Grimm, F., Kennedy, M. A., and Dahl, C. (2008) *Allochromatium vinosum* DsrC: solution-state NMR structure, redox properties, and interaction with DsrEFH, a protein essential for purple sulfur bacterial sulfur oxidation, *J. Mol. Biol.* **382**, 692-707.

Cramer, W. A., and Zhang, H. (2006) Consequences of the structure of the cytochrome b_6f complex for its charge transfer pathways, *Biochim. Biophys. Acta* **1757**, 339-345.

Cramer, W. A., Zhang, H., Yan, J., Kurisu, G., and Smith, J. L. (2006) Transmembrane traffic in the cytochrome b_6f complex, *Annu. Rev. Biochem.* **75**, 769-790.

Dagert, M., and Ehrlich, S. D. (1979) Prolonged incubation in calcium chloride improves the competence of *Escherichia coli* cells, *Gene* **6**, 23-28.

Dahl, C., Engels, S., Pott-Sperling, A. S., Schulte, A., Sander, J., Lübbe, Y., Deuster, O., and Brune, D. C. (2005) Novel genes of the *dsr* gene cluster and evidence for close interaction of Dsr proteins during sulfur oxidation in the phototrophic sulfur bacterium *Allochromatium vinosum*, *J. Bacteriol.* **187**, 1392-1404.

Dahl, C., Schulte, A., Stockdreher, Y., Hong, C., Grimm, F., Sander, J., Kim, R., Kim, S. H., and Shin, D. H. (2008) Structural and molecular genetic insight into a widespread sulfur oxidation pathway, *J. Mol. Biol.* **384**, 1287-1300.

Dambe, T., Quentmeier, A., Rother, D., Friedrich, C., and Scheidig, A. J. (2005) Structure of the cytochrome complex SoxXA of *Paracoccus pantotrophus*, a heme enzyme initiating chemotrophic sulfur oxidation, *J. Struct. Biol.* **152**, 229-234.

Dietrich, W., and Klimmek, O. (2002) The function of methyl-menaquinone-6 and polysulfide reductase membrane anchor (PsrC) in polysulfide respiration of *Wolinella succinogenes*, *Eur. J. Biochem.* **269**, 1086-1095.

Dubini, A., Pye, R. L., Jack, R. L., Palmer, T., and Sargent, F. (2002) How bacteria get energy from hydrogen: a genetic analysis of periplasmic hydrogen oxidation in *Escherichia coli*, *Int. J. Hydrogen Energy* **27**, 1413-1420.

Franz, B., Gehrke, T., Lichtenberg, H., Hormes, J., Dahl, C., and Prange, A. (2009) Unexpected extracellular and intracellular sulfur species during growth of *Allochromatium vinosum* with reduced sulfur compounds, *Microbiology* **155**, 2766-2774.

7 Reference List

Friedrich, C. G., Bardischewsky, F., Rother, D., Quentmeier, A., and Fischer, J. (2005) Prokaryotic sulfur oxidation, *Curr. Opin. Microbiol.* **8**, 253-259.

Frigaard, N.-U., and Dahl, C. (2009) Sulfur metabolism in phototrophic sulfur bacteria, *Adv. Microb. Physiol.* **54**, 103-200.

Gordon, E., Horsefield, R., Swarts, H. G. P., de Pont, J. J. H. H. M., Neutze, R., and Snijder, A. (2008) Effective high-throughput overproduction of membrane proteins in *Escherichia coli*, *Protein Expr. Purif.* **62**, 1-8.

Gräwert, T., Kaiser, J., Zepeck, F., Laupitz, R., Hecht, S., Amslinger, S., Schramek, N., Schleicher, E., Weber, S., Haslbeck, M., Buchner, J., Rieder, C., Arigoni, D., Bacher, A., Eisenreich, W., and Rohdich, F. (2004) IspH protein of *Escherichia coli*: studies on iron-sulfur cluster implementation and catalysis, *J. Am. Chem. Soc.* **126**, 12847-12855.

Grein, F., Venceslau, S. S., Schneider, L., Todorovic, S., Hildebrandt, P., Pereira, I. A. C., and Dahl, C. (2010) DsrJ, an essential part of the DsrMKJOP transmembrane complex in the purple sulfur bacterium *Allochromatium vinosum*, is an unusual triheme cytochrome *c*, *Biochemistry* **49**, 8290-8299.

Grimm, F., Cort, J. R., and Dahl, C. (2010a) DsrR, a novel IscA-like protein lacking iron- and Fe-S-binding functions, involved in the regulation of sulfur oxidation in *Allochromatium vinosum*, *J. Bacteriol.* **192**, 1652-1661.

Grimm, F., Dobler, N., and Dahl, C. (2010b) Regulation of dsr genes encoding proteins responsible for the oxidation of stored sulfur in Allochromatium vinosum, *Microbiology* **156**, 764-773.

Hamann, N., Bill, E., Shokes, J. E., Scott, R. A., Bennati, M., and Hedderich, R. (2009) The CCG-domain-containing subunit SdhE of succinate:quinone oxidoreductase from *Sulfolobus solfataricus* P2 binds a [4Fe-4S] cluster, *J. Biol. Inorg. Chem.* **14**, 457-470.

Hamann, N., Mander, G. J., Shokes, J. E., Scott, R. A., Bennati, M., and Hedderich, R. (2007) A cysteine-rich CCG domain contains a novel [4Fe-4S] cluster binding motif as deduced from studies with subunit B of heterodisulfide reductase from *Methanothermobacter marburgensis*, *Biochemistry* **46**, 12875-12885.

Hasler, J. A., Estabrook, R., Murray, M., Pikuleva, I., Waterman, M., Capdevila, J., Holla, V., Helvig, C., Falck, J. R., Farrell, G., Kaminsky, L. S., Spivack, S. D., Boitier, E., and Beaune, P. (1999) Human cytochromes P450, *Mol. Asp. Med.* **20**, 1-137.

Hedderich, R., Hamann, N., and Bennati, M. (2005) Heterodisulfide reductase from methanogenic archaea: a new catalytic role for an iron-sulfur cluster, *Biol. Chem.* **386**, 961-970.

Heiden, S., Hedderich, R., Setzke, E., and Thauer, R. K. (1993) Purification of a cytochrome *b* containing H$_2$:heterodisulfide oxidoreductase complex from membranes of *Methanosarcina barkeri*, *Eur. J. Biochem.* **213**, 529-535.

Heiden, S., Hedderich, R., Setzke, E., and Thauer, R. K. (1994) Purification of a two-subunit cytochrome-*b*-containing heterodisulfide reductase from methanol-grown *Methanosarcina barkeri*, *Eur. J. Biochem.* **221**, 855-861.

Hensen, D., Sperling, D., Trüper, H. G., Brune, D. C., and Dahl, C. (2006) Thiosulphate oxidation in the phototrophic sulphur bacterium *Allochromatium vinosum*, *Mol. Microbiol.* **62**, 794-810.

Hipp, W. M., Pott, A. S., Thum-Schmitz, N., Faath, I., Dahl, C., and Trüper, H. G. (1997) Towards the phylogeny of APS reductases and sirohaem sulfite reductases in sulfate-reducing and sulfur-oxidizing prokaryotes, *Microbiology* **143** 2891-2902.

Horton, R. M. (1995) PCR-mediated recombination and mutagenesis. SOEing together tailor-made genes, *Mol. Biotechnol.* **3**, 93-99.

Hussain, H., Grove, J., Griffiths, L., Busby, S., and Cole, J. (1994) A seven-gene operon essential for formate-dependent nitrite reduction to ammonia by enteric bacteria, *Mol. Microbiol.* **12**, 153-163.

Imhoff, J. F. (1984) Quinones of phototrophic purple bacteria, *FEMS Microbiol. Lett.* **25**, 85-89.

Imhoff, J. F. (2005) Genus II. *Allochromatium*, In *Bergey's Manual of Systematic Bacteriology* (Garrity, G., Brenner, D. J., Krieg, N. R., and Staley, J. T., Eds.), pp 12-14, Springer, New York.

Jormakka, M., Yokoyama, K., Yano, T., Tamakoshi, M., Akimoto, S., Shimamura, T., Curmi, P., and Iwata, S. (2008) Molecular mechanism of energy conservation in polysulfide respiration, *Nat. Struct. Mol. Biol.* **15**, 730-737.

Juncker, A. S., Willenbrock, H., Von Heijne, G., Brunak, S., Nielsen, H., and Krogh, A. (2003) Prediction of lipoprotein signal peptides in Gram-negative bacteria, *Protein Sci.* **12**, 1652-1662.

Kappler, U., Aguey-Zinsou, K.-F., Hanson, G. R., Bernhardt, P. V., and McEwan, A. G. (2004) Cytochrome c_{551} from *Starkeya novella*: characterization, spectroscopic properties, and phylogeny of a diheme protein of the SoxAX family, *J. Biol. Chem.* **279**, 6252-6260.

Kappler, U., Bernhardt, P. V., Kilmartin, J., Riley, M. J., Teschner, J., McKenzie, K. J., and Hanson, G. R. (2008) SoxAX cytochromes, a new type of heme copper protein involved in bacterial energy generation from sulfur compounds, *J. Biol. Chem.* **283**, 22206-22214.

Kelly, D. P., Shergill, J. K., Lu, W. P., and Wood, A. P. (1997) Oxidative metabolism of inorganic sulfur compounds by bacteria, *Antonie Van Leeuwenhoek* **71**, 95-107.

Keon, R. G., and Voordouw, G. (1996) Identification of the HmcF and topology of the HmcB subunit of the Hmc complex of *Desulfovibrio vulgaris*, *Anaerobe* **2**, 231-238.

Khoroshilova, N., Popescu, C., Munck, E., Beinert, H., and Kiley, P. J. (1997) Iron-sulfur cluster disassembly in the FNR protein of *Escherichia coli* by O_2: [4Fe-4S] to [2Fe-2S] conversion with loss of biological activity, *Proc. Natl. Acad. Sci. U. S. A.* **94**, 6087-6092.

Kovach, M. E., Elzer, P. H., Hill, D. S., Robertson, G. T., Farris, M. A., Roop, R. M., 2nd, and Peterson, K. M. (1995) Four new derivatives of the broad-host-range cloning vector pBBR1MCS, carrying different antibiotic-resistance cassettes, *Gene* **166**, 175-176.

Krafft, T., Gross, R., and Kröger, A. (1995) The function of *Wolinella succinogenes psr* genes in electron transport with polysulphide as the terminal electron acceptor, *Eur. J. Biochem.* **230**, 601-606.

Künkel, A., Vaupel, M., Heim, S., Thauer, R. K., and Hedderich, R. (1997) Heterodisulfide reductase from methanol-grown cells of *Methanosarcina barkeri* is not a flavoenzyme, *Eur. J. Biochem.* **244**, 226-234.

Laemmli, U. K. (1970) Cleavage of structural proteins during the assembly of the head of bacteriophage T4, *Nature* **227**, 680-685.

Lanciano, P., Magalon, A., Bertrand, P., Guigliarelli, B., and Grimaldi, S. (2007) High-stability semiquinone intermediate in nitrate reductase A (NarGHI) from *Escherichia coli* is located in a quinol oxidation site close to heme b_D, *Biochemistry* **46**, 5323-5329.

Larkin, M. A., Blackshields, G., Brown, N. P., Chenna, R., McGettigan, P. A., McWilliam, H., Valentin, F., Wallace, I. M., Wilm, A., Lopez, R., Thompson, J. D., Gibson, T. J., and Higgins, D. G. (2007) Clustal W and Clustal X version 2.0, *Bioinformatics* **23**, 2947-2948.

Laska, S., Lottspeich, F., and Kletzin, A. (2003) Membrane-bound hydrogenase and sulfur reductase of the hyperthermophilic and acidophilic archaeon *Acidianus ambivalens*, *Microbiology* **149**, 2357-2371.

Lemberg, R., and Barrett, J., (Eds.) (1973) *Cytochromes*, Academic Press, London.

Lemos, R. S., Fernandes, A. S., Pereira, M. M., Gomes, C. M., and Teixeira, M. (2002) Quinol:fumarate oxidoreductases and succinate:quinone oxidoreductases: phylogenetic relationships, metal centres and membrane attachment, *Biochim. Biophys. Acta* **1553**, 158-170.

Lemos, R. S., Gomes, C. M., and Teixeira, M. (2001) *Acidianus ambivalens* Complex II typifies a novel family of succinate dehydrogenases, *Biochem. Biophys. Res. Commun.* **281**, 141-150.

Li, W., Zhou, X., and Lu, P. (2004) Bottlenecks in the expression and secretion of heterologous proteins in *Bacillus subtilis*, *Res. Microbiol.* **155**, 605-610.

Lippard, S. J. (1995) *Bioanorganische Chemie*, Spektrum Akademischer Verlag, Heidelberg.

Londer, Y. Y., Pokkuluri, P. R., Tiede, D. M., and Schiffer, M. (2002) Production and preliminary characterization of a recombinant triheme cytochrome c_7 from *Geobacter sulfurreducens* in *Escherichia coli*, *Biochim. Biophys. Acta* **1554**, 202-211.

Lübbe, Y. (2005) Biochemische und molekularbiologische Untersuchungen zur Funktion von DsrN und DsrL im dissimilatorischen Schwefelstoffwechsel von *Allochromatium vinosum*, *PhD thesis*, University of Bonn.

Lübbe, Y. J., Youn, H.-S., Timkovich, R., and Dahl, C. (2006) Siro(haem)amide in *Allochromatium vinosum* and relevance of DsrL and DsrN, a homolog of cobyrinic acid a,c-diamide synthase, for sulphur oxidation, *FEMS Microbiol. Lett.* **261**, 194-202.

Magalon, A., Lemesle-Meunier, D., Rothery, R. A., Frixon, C., Weiner, J. H., and Blasco, F. (1997) Heme axial ligation by the highly conserved his residues in helix II of Cytochrome *b* (NarI) of *Escherichia coli* nitrate reductase A (NarGHI), *J. Biol. Chem.* **272**, 25652-25658.

Mander, G. J., Duin, E. C., Linder, D., Stetter, K. O., and Hedderich, R. (2002) Purification and characterization of a membrane-bound enzyme complex from the sulfate-reducing archaeon *Archaeoglobus fulgidus* related to heterodisulfide reductase from methanogenic archaea, *Eur. J. Biochem.* **269**, 1895-1904.

Mander, G. J., Pierik, A. J., Huber, H., and Hedderich, R. (2004) Two distinct heterodisulfide reductase-like enzymes in the sulfate-reducing archaeon *Archaeoglobus profundus*, *Eur. J. Biochem.* **271**, 1106-1116.

Maniatis, T., Fritsch, E. F., and Sambrook, J. (1982) *Molecular cloning : A laboratory manual* Cold Spring Harbor Laboratory, New York.

Martinis, S. A., Blanke, S. R., Hager, L. P., Sligar, S. G., Hoa, G. H., Rux, J. J., and Dawson, J. H. (1996) Probing the heme iron coordination structure of pressure-induced cytochrome P420cam, *Biochemistry* **35**, 14530-14536.

Marvin, K. A., Kerby, R. L., Youn, H., Roberts, G. P., and Burstyn, J. N. (2008) The transcription regulator RcoM-2 from *Burkholderia xenovorans* is a cysteine-ligated hemoprotein that undergoes a redox-mediated ligand switch, *Biochemistry* **47**, 9016-9028.

7 Reference List

Menon, N. K., Chatelus, C. Y., Dervartanian, M., Wendt, J. C., Shanmugam, K. T., Jr, H. D. P., and Przybyla, A. E. (1994) Cloning, sequencing, and mutational analysis of the *hyb* operon encoding *Escherichia coli* hydrogenase 2, *J. Bacteriol.* **176**, 4416-4423.

Miroux, B., and Walker, J. E. (1996) Over-production of proteins in *Escherichia coli*: mutant hosts that allow synthesis of some membrane proteins and globular proteins at high levels, *J. Mol. Biol.* **260**, 289-298.

Mohanty, A. K., and Wiener, M. C. (2004) Membrane protein expression and production: effects of polyhistidine tag length and position, *Protein Expr. Purif.* **33**, 311-325.

Nakajima, H., Honma, Y., Tawara, T., Kato, T., Park, S.-Y., Miyatake, H., Shiro, Y., and Aono, S. (2001) Redox properties and coordination structure of the heme in the CO-sensing transcriptional activator CooA, *J. Biol. Chem.* **276**, 7055-7061.

Nugent, T., and Jones, D. T. (2009) Transmembrane protein topology prediction using support vector machines, *BMC Bioinformatics* **10**, 159.

Ogawa, T., Furusawa, T., Nomura, R., Seo, D., Hosoya-Matsuda, N., Sakurai, H., and Inoue, K. (2008) SoxAX binding protein, a novel component of the thiosulfate-oxidizing multienzyme system in the green sulfur bacterium *Chlorobium tepidum*, *J. Bacteriol.* **190**, 6097-6110.

Oliveira, T. F., Vonrhein, C., Matias, P. M., Venceslau, S. S., Pereira, I. A. C., and Archer, M. (2008) The crystal structure of *Desulfovibrio vulgaris* dissimilatory sulfite reductase bound to DsrC provides novel insights into the mechanism of sulfate respiration, *J. Biol. Chem.* **283**, 34141-34149.

Page, C. C., Moser, C. C., and Dutton, P. L. (2003) Mechanism for electron transfer within and between proteins, *Curr. Opin. Chem. Biol.* **7**, 551-556.

Palmer, G. (2000) Electron Paramagnetic Resonance of Metalloproteins, In *Physical Methods in BioInorganic Chemistry* (Que, L., Ed.), pp 121-185, University Press Sausalito.

Paoli, M., Marles-Wright, J., and Smith, A. (2002) Structure-function relationships in heme-proteins, *DNA Cell Biol.* **21**, 271-280.

Pattaragulwanit, K., and Dahl, C. (1995) Development of a genetic system for a purple sulfur bacterium: Conjugative plasmid transfer in *Chromatium vinosum*, *Arch. Microbiol.* **164**, 217-222.

Pereira, I. (2007) Respiratory membrane complexes of *Desulfovibrio*, In *Microbial Sulfur Metabolism* (Dahl, C., and Friedrich, C. G., Eds.), pp 24-35, Springer, Berlin.

Pereira, I. A. C., and Xavier, A. V. (2005) Multi-heme *c* cytochromes and enzymes, In *Encyclopedia of Inorganic Chemistry* (King, R. B., Ed.), pp 3360-3376, Wiley, New York.

Pfennig, N., and Trüper, H. G. (1989) Anoxygeneic phototrophic bacteria, In *Bergey's manual of systematic bacteriology* (Staley, J. T., Bryant, M. P., Pfennig, N., and Holt, J. G., Eds.), pp 1635-1653, Williams & Wilkins, Baltimore.

Pfennig, N., and Trüper, H. G. (1992) The family *Chromatiaceae*, In *The Prokaryotes. A Handbook on the biology of bacteria: ecophysiology, isolation, identification, applications* (Balows, A., Trüper, H. G., Dworkin, M., Harder, W., and Schleifer, K.-H., Eds.), pp 3200-3221, Springer, New York.

Pierik, A. J., Duyvis, M. G., van Helvoort, J. M., Wolbert, R. B., and Hagen, W. R. (1992) The third subunit of desulfoviridin-type dissimilatory sulfite reductases, *Eur. J. Biochem.* **205**, 111-115.

Pires, R. H., Venceslau, S. S., Morais, F., Teixeira, M., Xavier, A. V., and Pereira, I. A. C. (2006) Characterization of the *Desulfovibrio desulfuricans* ATCC 27774 DsrMKJOP complex - a membrane-bound redox complex involved in the sulfate respiratory pathway, *Biochemistry* **45**, 249-262.

Pott-Sperling, A. S. (2000) Das *dsr*-Operon von *Allochromatium vinosum*: molekularbiologische Charakterisierung der Gene für die Schwefeloxidation, PhD thesis, University of Bonn.

Pott, A. S., and Dahl, C. (1998) Sirohaem sulfite reductase and other proteins encoded by genes at the *dsr* locus of *Chromatium vinosum* are involved in the oxidation of intracellular sulfur, *Microbiology* **144** 1881-1894.

Prange, A., Engelhardt, H., Trüper, H. G., and Dahl, C. (2004) The role of the sulfur globule proteins of *Allochromatium vinosum*: mutagenesis of the sulfur globule protein genes and expression studies by real-time RT-PCR, *Arch. Microbiol.* **182**, 165-174.

Ran, Y., Zhu, H., Liu, M., Fabian, M., Olson, J. S., Aranda, R., Phillips, G. N., Dooley, D. M., and Lei, B. (2007) Bis-methionine ligation to heme iron in the streptococcal cell surface protein Shp facilitates rapid hemin transfer to HtsA of the HtsABC transporter, *J. Biol. Chem.* **282**, 31380-31388.

Rath, A., Glibowicka, M., Nadeau, V. G., Chen, G., and Deber, C. M. (2009) Detergent binding explains anomalous SDS-PAGE migration of membrane proteins, *Proc. Natl. Acad. Sci.* **106**, 1760-1765.

Reedy, C. J., Elvekrog, M. M., and Gibney, B. R. (2008) Development of a heme protein structure electrochemical function database, *Nucl. Acids Res.* **36**, database issue D307-D313.

Reedy, C. J., and Gibney, B. R. (2004) Heme protein assemblies, *Chem. Rev.* **104**, 617-650.

Reijerse, E. J., Sommerhalter, M., Hellwig, P., Quentmeier, A., Rother, D., Laurich, C., Bothe, E., Lubitz, W., and Friedrich, C. G. (2007) The unusal redox centers of

SoxXA, a novel *c*-type heme-enzyme essential for chemotrophic sulfur-oxidation of *Paracoccus pantotrophus*, *Biochemistry* **46**, 7804-7810.

Rethmeier, J., Rabenstein, A., Langer, M., and Fischer, U. (1997) Detection of traces of oxidized and reduced sulfur compounds in small samples by combination of different high-performance liquid chromatography methods, *J. Chromatogr. A* **760**, 295-302.

Rieske, J. S., MacLennan, D. H., and Coleman, R. (1964) Isolation and properties of an iron-protein from the (reduced coenzyme Q)-cytochrome *c* reductase complex of the respiratory chain, *Biochem. Biophys. Res. Commun.* **15**, 338-344.

Rojas, N. R., Kamtekar, S., Simons, C. T., McLean, J. E., Vogel, K. M., Spiro, T. G., Farid, R. S., and Hecht, M. H. (1997) De novo heme proteins from designed combinatorial libraries, *Protein Sci.* **6**, 2512-2524.

Rother, D., and Friedrich, C. G. (2002) The cytochrome complex SoxXA of *Paracoccus pantotrophus* is produced in *Escherichia coli* and functional in the reconstituted sulfur-oxidizing enzyme system, *Biochim. Biophys. Acta* **1598**, 65-73.

Rothery, R. A., Blasco, F., Magalon, A., Asso, M., and Weiner, J. H. (1999) The hemes of *Escherichia coli* nitrate reductase A (NarGHI): potentiometric effects of inhibitor binding to NarI, *Biochemistry* **38**, 12747-12757.

Rothery, R. A., Blasco, F., Magalon, A., and Weiner, J. H. (2001) The diheme cytochrome b subunit (NarI) of *Escherichia coli* nitrate reductase A (NarGHI): structure, function, and interaction with quinols, *J. Mol. Microbiol. Biotechnol.* **3**, 273-283.

Rothery, R. A., Workun, G. J., and Weiner, J. H. (2008) The prokaryotic complex iron-sulfur molybdoenzyme family, *Biochim. Biophys. Acta* **1778**, 1897-1929.

Rubio, L. M., and Ludden, P. W. (2008) Biosynthesis of the iron-molybdenum cofactor of nitrogenase, *Annu. Rev. Microbiol.* **62**, 93-111.

Sander, J. (2005) Biochemische, molekularbiologische und bioinformatische Untersuchungen zum DsrMKJOP-Komplex von *Allochromatium vinosum*, *PhD thesis*, University of Bonn.

Sander, J., Engels-Schwarzlose, S., and Dahl, C. (2006) Importance of the DsrMKJOP complex for sulfur oxidation in *Allochromatium vinosum* and phylogenetic analysis of related complexes in other prokaryotes, *Arch. Microbiol.* **186**, 357-366.

Sapay, N., Guermeur, Y., and Deléage, G. (2006) Prediction of amphipathic in-plane membrane anchors in monotopic proteins using a SVM classifier, *BMC Bioinformatics* **7**, 255.

Saraiva, L. M., da Costa, P. N., Conte, C., Xavier, A. V., and LeGall, J. (2001) In the facultative sulphate/nitrate reducer *Desulfovibrio desulfuricans* ATCC 27774, the nine-haem cytochrome *c* is part of a membrane-bound redox complex mainly expressed in sulphate-grown cells, *Biochim. Biophys. Acta* **1520**, 63-70.

Schäfer, A., Tauch, A., Jager, W., Kalinowski, J., Thierbach, G., and Puhler, A. (1994) Small mobilizable multi-purpose cloning vectors derived from the *Escherichia coli* plasmids pK18 and pK19: selection of defined deletions in the chromosome of *Corynebacterium glutamicum*, *Gene* **145**, 69-73.

Schägger, H. (2006) Tricine-SDS-PAGE, *Nat. Protoc.* **1**, 16-22.

Schedel, M., Vanselow, M., and Trüper, H. G. (1979) Siroheme sulfite reductase isolated from *Chromatium vinosum*. Purification and investigation of some of its molecular and catalytic properties, *Arch. Microbiol.* **121**, 29-36.

Schiffer, A., Parey, K., Warkentin, E., Diederichs, K., Huber, H., Stetter, K. O., Kroneck, P. M., and Ermler, U. (2008) Structure of the dissimilatory sulfite reductase from the hyperthermophilic archaeon *Archaeoglobus fulgidus*, *J. Mol. Biol.* **379**, 1063-1074.

Schneider, L. (2007) Produktion eines ungewöhnlichen, purpurbakteriellen Trihäm-Cytochroms vom *c*-Typ in *Escherichia coli*, *Diplomathesis*, University of Bonn.

Schröder, I., Kröger, A., and Macy, J. M. (1988) Isolation of the sulphur reductase and reconstitution of the sulphur respiration of *Wolinella succinogenes*, *Arch. Microbiol.* **149**, 572-579.

Siebert, F., and Hildebrandt, P. (2008) *Vibrational spectroscopy in life science*, Wiley-VCH Verlag GmbH & Co KGaA, Weinheim.

Simianu, M., Murakami, E., Brewer, J. M., and Ragsdale, S. W. (1998) Purification and properties of the heme- and iron-sulfur-containing heterodisulfide reductase from *Methanosarcina thermophila*, *Biochemistry* **37**, 10027-10039.

Simon, J. (2002) Enzymology and bioenergetics of respiratory nitrite ammonification, *FEMS Microbiol. Rev.* **26**, 285-309.

Spiro, T. G. (1975) Resonance Raman spectroscopic studies of heme proteins, *Biochim. Biophys. Acta* **416**, 169-189.

Spiro, T. G., and Czernuszewicz, R. S. (1995) Resonance Raman spectroscopy of metalloproteins, *Methods Enzymol.* **246**, 416-460.

Spiro, T. G., and Strekas, T. C. (1974) Resonance Raman spectra of heme proteins. Effects of oxidation and spin state, *J. Am. Chem. Soc.* **96**, 338-345.

Steuber, J., Arendsen, A. F., Hagen, W. R., and Kroneck, P. M. (1995) Molecular properties of the dissimilatory sulfite reductase from *Desulfovibrio desulfuricans* (Essex) and comparison with the enzyme from *Desulfovibrio vulgaris* (Hildenborough), *Eur. J. Biochem.* **233**, 873-879.

7 Reference List

Steudel, R., Holdt, G., Visscher, P. T., and Gemerden, H. (1990) Search for polythionates in cultures of *Chromatium vinosum* after sulfide incubation, *Arch. Microbiol.* **153**, 432-437.

Thomas, P. E., Ryan, D., and Levin, W. (1976) An improved staining procedure for the detection of the peroxidase activity of cytochrome P-450 on sodium dodecyl sulfate polyacrylamide gels, *Anal. Biochem.* **75**, 168-176.

Todorovic, S., Jung, C., Hildebrandt, P., and Murgida, D. H. (2006) Conformational transitions and redox potential shifts of cytochrome P450 induced by immobilization, *J. Biol. Inorg. Chem.* **11**, 119-127.

Todorovic, S., Verissimo, A., Wisitruangsakul, N., Zebger, I., Hildebrandt, P., Pereira, M. M., Teixeira, M., and Murgida, D. H. (2008) SERR-spectroelectrochemical study of a cbb3 oxygen reductase in a biomimetic construct, *J. Phys. Chem. B* **112**, 16952-16959.

van Driessche, G., Devreese, B., Fitch, J. C., Meyer, T. E., Cusanovich, M. A., and Van Beeumen, J. J. (2006) GHP, a new *c*-type green heme protein from *Halochromatium salexigens* and other proteobacteria, *FEBS J.* **273**, 2801-2811.

van Wonderen, J. H., Knight, C., Oganesyan, V. S., George, S. J., Zumft, W. G., and Cheesman, M. R. (2007) Activation of the cytochrome cd_1 nitrite reductase from *Paracoccus pantotrophus*. Reaction of oxidized enzyme with substrate drives a ligand switch at heme *c*, *J. Biol. Chem.* **282**, 28207-28215.

Verte, F., Kostanjevecki, V., De Smet, L., Meyer, T. E., Cusanovich, M. A., and Van Beeumen, J. J. (2002) Identification of a thiosulfate utilization gene cluster from the green phototrophic bacterium *Chlorobium limicola*, *Biochemistry* **41**, 2932-2945.

von Jagow, E., and Schägger, H., (Eds.) (1994) *A practical guide to membrane protein purification*, Academic Press, San Diego, USA.

Wallace, C. J., and Clark-Lewis, I. (1992) Functional role of heme ligation in cytochrome *c*. Effects of replacement of methionine 80 with natural and non-natural residues by semisynthesis, *J. Biol. Chem.* **267**, 3852-3861.

Weaver, P. F., Wall, J. D., and Gest, H. (1975) Characterization of *Rhodopseudomonas capsulata*, *Arch. Microbiol.* **105**, 207-216.

Wollenberg, M., Berndt, C., Bill, E., Schwenn, J. D., and Seidler, A. (2003) A dimer of the FeS cluster biosynthesis protein IscA from cyanobacteria binds a [2Fe2S] cluster between two protomers and transfers it to [2Fe2S] and [4Fe4S] apo proteins, *Eur. J. Biochem.* **270**, 1662-1671.

Yamashita, T., Hoashi, Y., Watanabe, K., Tomisugi, Y., Ishikawa, Y., and Uno, T. (2004) Roles of heme axial ligands in the regulation of CO binding to CooA, *J. Biol. Chem.* **279**, 21394-21400.

7 Reference List

Yoshioka, S., Takahashi, S., Hori, H., Ishimori, K., and Morishima, I. (2001) Proximal cysteine residue is essential for the enzymatic activities of cytochrome P450cam, *Eur. J. Biochem.* **268**, 252-259.

Zehl, M., and Allmaier, G. (2004) Ultraviolet matrix-assisted laser desorption/ionization time-of-flight mass spectrometry of intact hemoglobin complex from whole human blood, *Rapid Commun. Mass Spectrom.* **18**, 1932-1938.

I want morebooks!

Buy your books fast and straightforward online - at one of world's fastest growing online book stores! Environmentally sound due to Print-on-Demand technologies.

Buy your books online at
www.morebooks.shop

Kaufen Sie Ihre Bücher schnell und unkompliziert online – auf einer der am schnellsten wachsenden Buchhandelsplattformen weltweit! Dank Print-On-Demand umwelt- und ressourcenschonend produziert.

Bücher schneller online kaufen
www.morebooks.shop

KS OmniScriptum Publishing
Brivibas gatve 197
LV-1039 Riga, Latvia
Telefax: +371 686 204 55

info@omniscriptum.com
www.omniscriptum.com

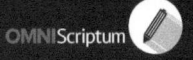

Printed by Books on Demand GmbH, Norderstedt / Germany